SCIENCE AND BIBLE CONVERGE

JULIANA JACOBS GRANT

GOD SAID

Isaiah 65:17 (NIV)

"See, I will create new heavens and a new earth. The former things will not be remembered, nor will they come to mind."

THIS IS NOT AN EXPRESSION OF HEAVEN AND EARTH.

THE NEW EARTH IS ORBITING

AMONGST THE PLANETS.

100 BILLION EARTH LIKE PLANETS,

NASA HAS DISCOVERED IT.

WHICH ONE IS YOURS?

PSALM 37:29 RIGHTEOUS SHALL INHERIT THE EARTH, AND DWELL UPON IT FOR EVER.

ARE YOU READY TO BELIEVE THE BIBLE?

READY OR NOT,

HE IS AT THE DOOR.

FROM THE AUTHOR

THE SLAP is a book that recounts the genesis of my calling into ministry by God. It evidently highlights the continuing work of God in using the simple to do great works. The book demonstrates that God is still in the business of using the 'world's reject' to bring about the manifestation of His glory.

(I have three versions of this book. I'm trying to reach everyone. THE SLAP is the first book, and the Holy Ghost continues to lead me to get into all the other markets. Since this is over 40 year of walking with a God who is kind to treat me as the prophet of old. I obeyed thus the version you are reading)

This book also speaks huge volume about the grace of God that brings salvation to everyone. Though God hates sin, yet He loves sinners, and He is always looking forward to our friendship and return to Him.

My story is proof that God's Word is true and has been inspired by God's Spirit. Through me, God has healed the sick, raised the dead, and performed powerful miracles. If He can do this, I can't imagine the extent to which he can use someone smarter, wiser, more intelligent, and strong like you.

I will leave you with this: we did not choose God; rather, He chose us, even when we knew Him not, by giving us His only son to die once and for all for the ultimate atonement of our sins. Although we live in a fallen world, if anyone would believe and accept Him and walk in His precepts, assuredly, his hope of eternal life has just been guaranteed.

Therefore, this is not a book to condemn anyone or make someone appear holier than others. Rather, it's a divine exposition of how God's grace can use an ordinary man to do the extraordinary. Just believe

God bless you. Amen, Juliana Jacobs Grant.

"Lord, are you at this time going to restore the kingdom to Israel?"

He said to them: “It is not for you to know the times or dates the Father has set by his own authority. But you will receive power when the Holy Spirit comes on you; and you will be my witnesses in Jerusalem, and in all Judea and Samaria, and to the ends of the earth.” (Acts 1: 6-8, NIV)

Often, we worry about our little world and the things we know but God has a bigger plan. He told the disciples to think big. **THINK BEYOND ISRAEL**. Witness for me from here to the whole world.

Behold love,

Jeremiah 29:11 (KJV): “For I know the thoughts that I think toward you, saith the LORD, thoughts of peace, and not of evil, to give you an expected end.”

Isaiah 64:4 (KJV): “For since the beginning of the world men have not heard, nor perceived by the ear, neither hath the eye seen, O God, beside thee, what he hath prepared for him that waited for him.”

PREFACE

God's passionate love for humanity

But one testified in a certain place, saying: 'What is man that You are mindful of him, Or the son of man that You take care of him? You have made him a little lower than the angels; You have crowned him with glory and honor, And set him over the works of Your hands. You have put all things in subjection under his feet' – Hebrews 2:6-8 (NKJV).

In the beginning of creation, God made man in His very image – to look like Him, walk like Him, talk like Him, create like He does and give physical expression to God's manifold attributes. If you were given an opportunity to see Adam as God created him, you would see how much he looked and acted like his Creator. He resembled God so much that the devil got envious and decided to upset the beautiful relationship between God and the first man, Adam. God has a special love for humans that made even the angels of heaven wonder in amazement at how the Creator could give us such a special place in His heart.

We are created in God's very image and given the special privilege to rule the earth – this is exactly why we are given dominion over the earth and all that is in it (See Genesis 1:28).

God's many gifts for humanity

While God's greatest gift to the world is His Son, who redeems us from the fall and its effects (John 3:16), He still has many other gifts us for us. There are different kinds of gifts in the Bible; some of which are:

a) The gift of Persons – Jesus being the greatest of them, the Holy Spirit our helper and comforter, and the angels who watch over God's people (John 1:12,13 Acts 2:38 Hebrews 1:14).

b) The gift of eternal life which we receive at salvation (See Romans 6:23).

c) The gifts/manifestations of the Holy Spirit (See 1 Corinthians 12:4-11).

d) The gift of physical things that God made for man to use, enjoy and explore. Some of these gifts were created even before humans. These gifts include the animal world, the world of plants and flowers with lovely fragrances, food and water, the moon, the sun, the air that we breathe, the stars, the planetary bodies and the entire universe itself (See Genesis 2:1, Acts 14:17).

The gifts of Persons, eternal life, and manifestations of the Spirit are exclusively for God's own people and can be accessed by faith in Christ the Son of God – but the gifts of physical things are open and available to the entire human race, both the saints of the Lord and the sinners of the devil. 'For He is kind to the unthankful and evil.' (Luke 6:35).

This explains why science is exploring the world in outer space – studying the stars, the moon, the sun and other planetary bodies. NASA planet hunters recently discovered an earth-sized planet in the habitable zone. The planet, named "TOI 700 d" is relatively close to earth – about a 100 light years away – and the discovery was confirmed by the Spitzer Space Telescope. The scientists might think that they are achieving all their feats because they are smart and intelligent on their own, but the truth is that the Creator-God had always wanted men to discover and marvel at His works! The unbelieving but intellectually wise men are permitted to study, observe and research the heavens all they want, but God has placed limitations on their achievements. "When I made the clouds its garment, and thick darkness its swaddling band; When I fixed My limit for it, and set bars and doors; When I said, 'This far you may come, but no farther, and here your proud waves must stop!'" – Job 38:9-11.

The intruders and the heirs

'The heavens declare the glory of God; And the firmament shows His handiwork.'

– Psalm 19:1.

Just like a loving father would provide good learning materials and amusing toys for the education of his child, God has created so many things – both in heavenly places and the earth – for the learning and amusement of His children. The heavens and all the beauty that God displays in it weren't created for the devil and those who rebel against God – but for His children who love Him in return and seek to serve Him with all their hearts.

'Lift up your eyes on high and see: who created these? He who brings out their host by number, calling them all by name, by the greatness of his might, and because he is strong in power not one is missing.' – Isaiah 40:26

While the saints of God may not have the physical or technological resources to explore outer space, God has given us something greater than what scientists and researchers could ever hope to find. While they observe and study the stars and all that's out there, they cannot influence or control them. But God has given us the power and authority to influence the stars and the planetary bodies! The Bible talks about Jesus our King coming in the clouds of heaven (Revelation 1:7), and holding bright and shiny starts in His right hand (Revelation 1:16).

Jesus is the head and we the church are His body. He shares His authority with us, making it possible – through the Holy Spirit – to also influence outer space and affect the behavior of heavenly bodies. Listen to what the Lord is saying right now: 'The effective and united prayers and praises of the saints of God will always produce tremendous power through the Holy Spirit to change the course and direction of the stars and planetary bodies – which in turn have effects on the earth.' Science might never be able to prove this – but it is true!

The earnest prayers of Elijah affected the planetary bodies and the stars and caused rain to fall (James 5:18, 1 Kings 18:41-45). The Bible also speaks of God's two witnesses who had the power to shut heaven so that no rain could fall (Revelation 11:6). When our King was crucified, the stars, the moon, and the sun all withdrew their light causing great darkness on the earth – they simply could not behold the suffering of their Creator and ran to hide as a sign of honor, respect and reverence for the King (See Luke 23:44,45).

Jesus: God's Key to knowledge

'And attaining to all riches of the full assurance of understanding, to the knowledge of the mystery of God, both of the Father and of Christ, in whom are hidden all the treasures of wisdom and knowledge.'

– Colossians 2:2,3.

There is nothing education or science will discover that is not written in the Word of God in some fashion. For example, the Bible records how God created the earth and made it hang on nothing thousands of years before science made that discovery! 'He stretches out the north over empty space; He hangs the earth on nothing.' – Job 26:7. The Bible also mentions the importance of hand washing in running water (See Leviticus 15:13), and even the importance of blood in our lives (See Leviticus 17:14). You need to understand that the Bible is the first place to go for answers – whether you're conducting research or simply studying God's wonderful creation!

No scientist or researcher can ever fully understand God's creation without Christ Jesus. He alone is the key access to God's vast realm of knowledge and wisdom, and through the Holy Spirit, He will teach you all the things that you really need to know if you open your heart and mind to Him (See John 14:26).

We cannot fully comprehend God's creation through the knowledge we acquire in schools – His universe is bigger than our minds can take! He designed creation – with its amazing creatures, beautiful colors, sounds and fragrances – for His pleasure and glory (See Revelation 4:11). If we are ever going to understand creation, we must depend on the Lord and ask Him to teach us the truth about His purposes for creating the stars of heaven, the moon, the sun, and the planets.

However, it is more important that we first find and discover our God-ordained purpose for our existence before we get to discover the divine purpose for the heavenly bodies. I understand it might be difficult to believe this – but every star in the sky has a purpose for which Jesus created it and He calls each one of them by name! Here's how the Bible put it: 'He counts the number of the stars; He calls them all by name. Great is our Lord, and mighty in power; His understanding is infinite.' – Psalm 147:4, 5.

New creation for God's people

'But the day of the Lord will come as a thief in the night, in which the heavens will pass away with a great noise, and the elements will melt with fervent heat; both the earth and the works that are in it will be burned up. Therefore, since all these things will be dissolved, what manner of persons ought you to be in holy conduct and godliness, looking for and hastening the coming of the day of God, because of which the heavens will be dissolved, being on fire, and the elements will melt with fervent heat? Nevertheless we, according to His promise, look for new heavens and a new earth in which righteousness dwells.'

– 2 Peter 3:10-13.

As wonderful and enlightening as science's discoveries have been, the scriptures tell us that God will do away with the whole earth, heavens, and everything in them – including the sun, moon, and stars – and will create a new earth, heaven, and the host of them. The earth and everything in it – except the saints – are corrupted by sin, wickedness, violence, bloodshed and the presence/activities of demon spirits.

'See, I will create new heavens and a new earth. The former things will not be remembered, nor will they come to mind.' – Isaiah 65:17.

Jesus is going to give His people great power over the new earth, allowing us to rule, influence and direct creation – fully restoring back to us that which Adam lost in the garden! However, God does not desire this for only a few people, but for 'as many as received Him' (John 1:12). Are you a scientist or an astronaut who does not believe? Are you aware that this earth and the starry heavens will one day be destroyed? I tell you – God is real, and He will make Himself real to you if you give Him half a chance!

If you're reading this and you have yet to receive Christ into your heart and make Him Lord of your life, I urge you to do so right now. Through Jesus, God offers you forgiveness and deliverance from the pulling-power of sin, but it takes your faith and words of commitment to accept what God offers. Believe that Jesus shed His blood for you – pray from your heart right now – and consciously say these words: Jesus, I come to You and I ask that You forgive me of all my sins, cleanse me in and out with your precious blood, empower me by the Holy Spirit to live a sin-free life, and teach me about the things of God. Come into my heart, life and home and become the Lord of my life and destiny. Amen!

Contents

INTRODUCTION

I am not writing this book for you to know my story but for you to instead consider that we live in a fallen world. This implies that other beings occupied the earth before humans were created.

Here, the two beings to be cautious about are the sons of God and the spirits of God. In Job 38:4-7, the conversation was about the beginning of the earth and both the sons of God, which some believe are angels, and the spirits of God. I will only speak a little about these two in this book and elaborate in another book in more details. I want this book to have an immediate impact on readers. I want to either give readers the opportunity to be set free immediately, to set someone free immediately or to say "God, I love you with all my heart; I know when I need someone, you will be available". Trust that God is your Father and will do good through and with you.

Job 38:1-7: "Then the LORD answered Job out of the whirlwind, and said, who is this that darkeneth counsel by words without knowledge? Gird up now thy loins like a man; for I will demand of thee, and answer thou me. Where wast thou when I laid the foundations of the earth? Declare, if thou hast understanding. Who hath laid the measures thereof, if thou knowest? Or who hath stretched the line upon it? Whereupon are the foundations thereof fastened? Or who laid the corner stone thereof; when the morning stars sang together, and all the sons of God shouted for joy?"

Hebrews 1:7: "In speaking of the angels he says, 'He makes his angels spirits, and his servants flames of fire.'"

Psalm 104:4: "He makes the winds His messengers, flames of fire His servants."

Genesis 1:2: "Now the earth was formless and empty, darkness was over the surface of the deep, and the Spirit of God was hovering over the waters."

Job 1:6: "One day the angels came to present themselves before the Lord, and Satan also came with them."

Satan was an archangel that previously lived in heaven and served God. However, pride entered his heart and his rebellion caused him, along with his rebellious allies, to fall from heaven.

Jude 1:6: "And the angels who did not keep their positions of authority but abandoned their proper dwelling—these he has kept in darkness, bound with everlasting chains for judgment on the great Day."

Though people were pure in the beginning of time, sin and a desire to be like God also brought corruption. Genesis 3:22 says, "And the Lord God said, Behold, the man is become as one of us, to know good and evil: and now, lest he put forth his hand, and take also of the tree of life, and eat, and live forever". We were never meant to die spiritually, but we sinned. Adam and Eve were kicked out of the garden, and sin was present in all future generations.

The angels that fell with Satan took some humans as wives.

Genesis 6:1-2: "When human beings began to increase in number on the earth and daughters were born to them, the sons of God saw that the daughters of humans were beautiful, and they married any of them they chose."

The revelations in this book are heavy and hard to pen. Demons can be identified as Chaldeans. They were previously God's children because He created them as angels. However, after they were cast out for coinciding with Satan's pride, they became evil. I recommend listening to these sermons by Pastor Robert

Morris: **https://sermons.love/robert-morris-sermons/free-indeed/3565-robert-morris-beware-of-the-chaldeans.html** and **https://www.gotquestions.org/Chaldeans.html**.

No one can stop evil from being evil. Satan has an unrelenting malicious plan for humans, fueled by jealousy. He wants to stop humanity from becoming the apple of God's eye by deterring people from the path of salvation. Now, hell is increasing itself to receive man.

Isaiah 5:14: "Therefore hell hath enlarged herself, and opened her mouth without measure: and their glory, and their multitude, and their pomp, and he that rejoiceth, shall descend into it."

The only escape from eternal torture can be found in Romans 10:13: "For whosoever shall call upon the name of the Lord shall be saved."

Hebrews 2:3: "How shall we escape if we ignore so great a salvation? This salvation, which was first announced by the Lord, was confirmed to us by those who heard him."

John 1:12: "Yet to all who did receive him, to those who believed in his name, he gave the right to become children of God—"

Romans 8:1-3: "Therefore, there is now no condemnation for those who are in Christ Jesus, because through Christ Jesus the law of the Spirit who gives life has set you free from the law of sin and death. For what the law was powerless to do because it was weakened by the flesh, God did by sending his own Son in the likeness of sinful flesh to be a sin offering. And so he condemned sin in the flesh."

Created to rule

After salvation, your life needs to be transformed to the new life offered by Christ. Learn to think in a new way. Learn to be obedient to God. Learn to fight the good fight as good soldiers.

Here are two things that redeemed people must know:

1. Fight the good fight of faith. Stand firm in your belief. While Jesus Christ does redeem us through grace, it is

important for us to work to carry out his purpose and destroy the works of Satan. This has been God's expectation from the beginning.

2. The overcoming help of the Holy Spirit is poured on the church and we now have power to cast the devil out. The devil dwells and works through disobedient people, people we once were. When they die, the Lord will not receive their spirit because it is evil, and the time has not come for them to be locked for good. They will simply be spirits without bodies looking for someone to possess.

The Bible equips us with the full armor of God in Ephesians 6:

"10 Finally, be strong in the Lord and in his mighty power. 11 Put on the full armor of God, so that you can take your stand against the devil's schemes. 12 For our struggle is not against flesh and blood, but against the rulers, against the authorities, against the powers of this dark world and against the spiritual forces of evil in the heavenly realms. 13 Therefore put on the full armor of God, so that when the day of evil comes, you may be able to stand your ground, and after you have done everything, to stand. 14 Stand firm then, with the belt of truth buckled around your waist, with the breastplate of righteousness in place, 15 and with your feet fitted with the readiness that comes from the gospel of peace. 16 In addition to all this, take up the shield of faith, with which you can extinguish all the flaming arrows of the evil one. 17 Take the helmet of salvation and the sword of the Spirit, which is the word of God."

Cover yourself with the Word of God, by which Jesus overcame the temptation of the devil.

Habakkuk 2:14: "For the earth will be filled with the knowledge of the glory of the Lord as the waters cover the sea."

Isaiah 11:9: "They will neither harm nor destroy on all my holy mountain, for the earth will be filled with the knowledge of the Lord as the waters cover the sea."

Testimonials will come from some tests you endure. The stories you will gain and characteristics you will build will seem to come from the plot of the enemy to prove you as unqualified to be an heir of Christ. However, God works everything out for our good, and as His child, we can endure walking through the valley of the shadow of death and emerge as new creations in Christ capable of managing God created for us.

Genesis 1:26: "And God said, Let us make man in our image, after our likeness: and let them have dominion over the fish of the sea, and over the fowl of the air, and over the cattle, and over all the earth, and over every creeping thing that creepeth upon the earth."

I hope these scriptures will convince you to believe that both the spirits and the sons of God were on earth before the creation of man. The existing children of God at that time were giants which may just have been created to house the spirits but they were not made in the likeness of God. I believe they were made in God's image but not likeness. The Bible says they loved women and took them for wives against God's wishes in Genesis 6:1, As shown above.

2 Peter 2:4: "For if God did not spare angels when they sinned, but cast them into hell and committed them to chains of gloomy darkness to be kept until the judgment;"

Isaiah 14:12: "'How you are fallen from heaven, O Day Star, son of Dawn! How you are cut down to the ground, you who laid the nations low!'"

Corruption on the Earth

After the creation of man in Genesis, humans became increasingly wicked. God decided to wipe out the earth and create a generation of people that may choose him by drowning all men except Noah's family who took refuge in the ark in obedience to God.

After the flood, God still saw an incomplete resolution and promised that He would no longer use water to punish men. Rather, He'd use brimstone and fire. He later used brimstone and fire on Sodom and Gomorrah for their wickedness, lack of generosity and self-absolving behaviors.

How are young men and woman supposed to know about these ancient spirits who have lived many years before they existed? That seems unfair, like throwing a child in a lion's den and wondering why he did not make it.

God is just asking one thing from us: to believe him and trust Him

The reason God is frustrated is because of our lack of acknowledgement that He is the way, the truth, and the life. He wants us to cling to him. He kicked out Satan and all the Children of disobedience. If they didn't disobey God, they would have been with him. Once we are with him, we are safe. Have hard confidence in him more than you trust your friends, mate, and family. When you are aware of how good your family is, you brag about them. Well, brag about your God to your problems! Speak to evil spirits as though you know you belong to God and with authority cast out all evil spirits.

Luke 10:19: I have given you authority to trample on snakes and scorpions and to overcome all the power of the enemy; nothing will harm you. They will not harm you.

When God wants to do something, He uses humans. Conversely, when Satan wants to do something, he needs humans. God is good and does not force himself on people. Ask, Seek, and Knock – Matthew 7:6-8 says:

Ask and it will be given to you; seek and you will find; knock and the door will be opened to you. For everyone who asks receives; he who seeks finds; and to him who knocks, the door will be opened.

God, the maker of everything on earth and in heaven who created the evil and the good, says He has given you authority; believe it! A city gives a police officer a badge and that gives him the right to arrest someone and they believe in it. A policeman goes to work every day knowing his authority and everyone obeys him. You now have the same right to police this universe and serve God. Believe God.

I am adding my faith to yours. A prophecy came that this book has the hand of God on it and I receive it because God commanded me to write this. I feel He has written it Himself because I am not a wordsmith. The Bible says believe the prophet that you will prosper; believe the Lord and you will be established.

2 Chronicles 20:20: "Early in the morning they left for the Desert of Tekoa. As they set out, Jehoshaphat stood and said, 'Listen to me, Judah and people of Jerusalem! Have faith in the LORD your God and you will be upheld; have faith in his prophets and you will be successful.'"

I agree with you that any prayer you pray in the will of God must come to pass now, as it is written that if two agree on a thing, it shall come to pass in Jesus' name. Amen.

Matthew 18:19: "Again, I tell you truly that if two of you on the earth agree about anything you ask for, it will be done for you by My Father in heaven."

Matthew 21:22: "If you believe, you will receive whatever you ask in prayer."

Mark 11:24: "Therefore, I tell you, whatever you ask in prayer, believe that you have received it, and it will be yours."

INTRODUCTION TO ESCHATHOLOGY

Walking with God has always had an end goal that the follower is not only blessed now and on this side of life, but also receives something more eternal than the earthly blessings that He is well and able to bless you with.

God's initial plan and purpose for creating man was for Him to have companions who would understand Him, look like Him in image and likeness, and fulfill His mandate on the earth. God wanted to make a being that would not only serve Him, but fellowship with Him and have the same mentality that God had toward everything. God made man, not just to worship Him, but to accomplish of dominion and inheritance that God had planned for man.

God was always committed to this dream and would not let man be in the dark about His love and plan. This the Psalmist understood by study. He said in Psalms 8:4, "What is man that you are mindful of him, and the son of man that you care for him?" If you are still asking this question, then it is a direct consequence of your ignorance to the plan of God that would eventually make God come to earth in the form of man to die and be resurrected again. All this for what? Just so you can go to heaven and escape hell? No sir, it stretches far beyond that.

Do you know that there are rewards for walking with God? I am not talking of simply protection from evil or abundance, so that you won't lack food, clothes, cars, houses, etc. These are all good, but they're merely a scratch of the things that God has planned as rewards for those who are and will keep walking with Him.

Do you know the extent of the earth and the heavens that God created? The Bible says God made the "heavens" in Genesis

1:1. What are these heavens and where are they? I believe they are not just the spiritual heavens, but the cosmos and the entirety of the multiverse that God created.

Scientists have done well to bring to our awareness the existence of planets and worlds that we would have never known about in the past. There are about six billion earth-like planets or more, that scientists have discovered in their searches. I believe God is revealing all these to humans, so that we know that there is a whole lot that He has made, and we don't even know the least of it yet.

For many years to come, we will continue to discover new things that God already made from the beginning; things that reveal the true dimensions of His great and abundant creations. You must understand that it is virtually impossible to really comprehend all that God has made from this standpoint of reality.

Among all these things that God created, do you remember why He made man and what He told man while blessing him?

God said to man in Genesis 1:28, "Be fruitful and multiply and fill the earth and subdue it, and have dominion over the fish of the sea and over the birds of the heavens and over every living thing that moves on the earth." This was the original order of things that Jesus had to come and restore through the pouring of His blood.

There is an inheritance for every man that has been "born-again" and received back the authority that God gave in the beginning. If you are in Christ, there is hope for glory according to Colossians 1:27. The glory God has planned, will partly manifest as a reward to those who followed Him. We will inherit the earth and the things that God has created, all the worlds and the billions of planets that man does not even know about yet. There is so much glory to be inherited from God our Father when Jesus returns again. So, for the believer, the second coming is not a dreadful thought, but one that fills him with joy and hopeful expectation of what is his inheritance. Hallelujah!

God says in Psalms 37:29, "The righteous shall inherit the land and dwell upon it forever." Jesus went to prepare a place for the believers who will be welcomed into God's eternal presence of glory and comfort. Remember the parable of the servants and the talents? Jesus said at the end that whoever has worked and gained some will be given more.

There is going to be a reward day, and we will inherit much more than we presently have, only if we walk faithfully to the end.

Blessed be the God and father of our Lord Jesus Christ, now and forever in every place. Amen.

Chapter 1

THE SLAP (REVIVAL STARTS)

What was God up to?

I had nothing on my mind but to go to school and get straight A's. However, God had another plan and his Word confirmed it according to Jeremiah 29:11:

"For I know the plans I have for you," declares the LORD, "plans to prosper you and not to harm you, plans to give you hope and a future".

Isaiah 55: 8-9 is another word to consolidate God's plan for me:

For my thoughts are not your thoughts, neither are your ways my ways, saith the LORD. For as the heavens are higher than the earth, so are my ways higher than your ways, and my thoughts than your thoughts.

I was on the ninth cloud because God had to speak to me again. I thought He was reassuring me. However, I have since learned that when God is reassuring you, it's not a time to celebrate but a time to be cautious.

God doesn't waste time. So, as that was happening, the school patron came to assign me to my bed and the welcoming committee came to tell me about the school. While I waited to be assisted, the driver was still much available. The seniors were informed that they got first choice as to where they prefer to stay. The same also applied to me but all I needed was just a quiet corner.

"I'm a Christian," I said, but I didn't know why I said this. I didn't have to answer that way.

"Hmmm, you are one of those…", one of the girls standing to carry my luggage and to welcome me said. While she was still talking, I lifted my hand and slapped her. I have never done that before nor have I done it again. I don't slap people. The only people I lay hands on are my brothers. All the time we have fought, I won. Of course, they are not supposed to touch a lady. Unfortunately, now, I have slapped a stranger. I wondered what in the world I had done; I must have lost my mind. It was like someone lifted my hand and slapped this young woman. What in the world just happened? Everyone at the scene was shocked and started screaming. This got the attention of about five hundred students and I have no idea how to get out of the situation.

I didn't know why the young woman opened her mouth and I didn't know why I slapped her. A peacemaker turned a fighter. I told her that I was a Christian one second and slapped her the other. This is quite heart-breaking.

How could I explain this? My God wouldn't do that. Although I have an affluent family and I presumed I had protection from them, I have been told never to start a fight, but I do not retreat from a fight started against me. My mother paid a lot of money to send me to Queens School, so I can carry myself with dignity as a princess. This is one of the reasons why my mother and some family members called me such an endearing term. This situation negates every expectation from my family and friends but more importantly from me who confessed a higher moral standard of loving Jesus – a believer.

The girl I slapped was the principal's daughter. Her family was responsible for the school of Christianity. They used to spank Bible study students and the children. It was a dark place. I was still on the scene of announcing that I'm a Christian. The students looked at me with expressions that said "You do not know who you just slapped". I was a spoiled Christian child without a care for anything, except the abuse and beatings I'd receive when I went to Bible study or crusaded in rebellion to my upbringing. My family did not get my born-again status and how I longed to feed my spirit. I looked at the students from head to toe like nothing didn't matter to me whatsoever. I didn't saya word that I shouldn't say. I only mattered to Jesus. He was the only one who cared about my behavior, humility and kindness.

I felt I had let God down and was ashamed of myself. I was shocked that this violence was launched by me. I didn't want them to know that I was afraid, so I stood my ground. First, I was a child of God. Second, physically, if you remove that equation, I came from a family that would defend me. Either way, I was covered.. I stayed busy thinking of what I was going to say and how to correct this mistake.

The school matron defended me and said, "No! Julie did not do that". She remembered me from the Oral Roberts Crusade. During the crusade, I laid hands on hundreds of people, her son included. Her son was mentally ill and he was healed instantly. The matron remembered me, and she was convinced that I was not the person that slapped the student. She knew that I was of the light as it is written in Matthew 5:14-16:

"Ye are the light of the world. A city that is set on a hill cannot be hid. Neither do men light a candle, and put it under a bushel, but on a candlestick; and it giveth light unto all that are in the house. Let your light so shine before men that they may see your good works, and glorify your Father which is in heaven"

Though I was happy that she defended me, in my mind, I knew I did it.

I didn't know how she knew me at the time, but I needed someone to defend me and that's all I wanted. I was in a mess and I needed this to be cleaned up as soon as possible.

"Do you remember me?", the matron inquired.

"No", I said.

"My son was sick and she was the one who prayed for him. Now, he is healed," she said to the group.

I didn't remember healing her son, but I remembered going to the Oral Roberts Ministry Evangelistic Outreach. For me, it was a risky outreach to attend because had my mother knew I attended, she would have beaten me. Amusingly, I took beatings for Jesus on a regular basis because I was stubborn.

The outreach put thousands of people in line to be prayed for. It was requested of us to come forth and pray for the sick. While I was praying for the matron's son, the demon in him spoke before he left and said in a terrifying voice, "You chased me out of my house; you wait, and I will show you". I commanded the demon to get out in Jesus Name. Lo and behold, the demon headed out and the boy was set free.

Chapter 2

BACKLASH

When I returned home, my mother beat me and kicked me out of the house. Cryptically, the same spirit that left the matron's son jumped into her. The next day, while I was sitting in front of the house, she came and said, "Who did that to you"? My brother replied that she did it. I was terribly bleeding and full of sores. Notwithstanding, I tried to remember the matron's son, but I couldn't because I prayed for about 300 people during the outreach.

The matron told everyone how his son was healed, and they listened. He had not finished saying this when one of the girls collapsed and started convulsing – she was having an epileptic seizure. I said they should excuse me while they called for an ambulance. I headed to where she was. Prior to this, all week long, the Holy Ghost had laid this profession on my lips:

"Jesus who your fathers crucified is still going about doing good".

This is found in Acts 2:36 and it says, "Therefore let all the house of Israel know assuredly, that God hath made the same Jesus, whom ye have crucified, both Lord and Christ."

The Word that came to mind was Acts 10:38, "How God anointed Jesus of Nazareth with the Holy Ghost and with power: who went about doing good and healing all that were oppressed of the devil; for God was with him". Taking confidence from confessing these words all week long, I spoke and commanded this foul spirit to let her go and never to come back in the name of Jesus.

In a jiffy, the demon was gone. This is indeed God's power. Some of the people that experienced this brought their rosaries, but I requested she should be left alone.

The girls treated me like a hero and people were curious to know how possible that could be. A few moments ago, I was in a fight, but all of a sudden, God turned it around to bring about the salvation of his people. Beyond doubt, God wanted me to know he was with me. My burden was light. They kept on asking how I did it. I talked to them that it was Jesus whom your fathers crucified; he is still going about doing good. I told them to sing and praise God, worship him and their sins would be forgiven. I said, "Worship Him; He still loves you. Worship Him, He still thinks about you".

I began to teach them that there is no wrong or evil thing they have done that Jesus hasn't paid for. I asked them to worship and give their life to him if they are to receive from him. We had a worship session and later sent them back their churches to proclaim their salvation. I also requested that they should be discipled in their respective local churches. They said, "No ... we don't have churches. There is no church or a Bible study center around here". This was because the principal stopped the spreading of Christianity in the region.

The principal is the father of the girl I slapped. I wrestled within myself to know why I slapped her, but I realized the incident lines up with the word of God, as it is written in Matthew 11:12 which says, "And from the days of John the Baptist until now, the kingdom of heaven suffered violence and the violent take it by force".

This principal had scattered the Bible study and the cry of God's children had made it to heaven. Psalms 8:2 says, "Out of the mouth of babes and suckling's hast thou ordained strength because of thine enemies, that thou mightiest still the enemy and the avenger." Matthew 21:16 says, "And said unto him, Hearest thou what these say? And Jesus saith unto them, Yea; have ye never read, Out of the mouth of babes and suckling's thou hast perfected praise?"

I am strange spirit-filled student – now I've become the instrument that God has chosen. Candidly, when I gave my life to Christ, I would not have imagined doing anything substantial but from that time onward. Now, I have found myself as God's secret weapon.

God Was Offended

The spirit of the Lord was offended by the principal's family's rebellious act to the kingdom of God in the region of Africa. "Now, it's time to start a fellowship," I said audaciously. After lunch the next day, we met and the school authorities scheduled a meeting for the upcoming Monday to discuss the school's policies on school-related violence. The meeting showcased many of the students who got saved. They were Christians who weren't allowed to pray.

At this point, God showed me that He is God. Then, I was a child who was bewildered by the matter. All I knew was that God had been the driving force, but I could neither see nor understand it. Having encountered God, I prayed to Him and asked for the opening and filling of my mouth with His words. Since I did not know what was going on, I just asked God to show the principal that He is in control of the school.

Meeting with the Principal

I met with the Principal on the second floor of his home. He rocked back and forth in his chair. He stared straight at me and said, "You are the student who slapped my daughter?"

"Your daughter insulted me," I replied. He asked whose child I was. I told him my auntie was the Commissioner. "Oh …ok," he said.

God Showed Up

A shooting star then fell from the heavens and aimed directly at the principal's campus home. I had never seen anything like it before in my life. **The shooting star** came and fizzled in front ofhis lawn.

"What is that?", he said frighteningly.

The Lord spoke out of my mouth, "That is just one of the servants of God".

And so as it was written in Mark 13:25, "...the stars of heaven shall fall, and the powers that are in heaven shall be shaken."

He probably thought that I was invoking powers from heaven to get him, but I'd never seen such a sign. The Holy Ghost had spoken but I couldn't believe what God was saying: a shooting star is a servant of God!

"I heard that you are trying to host a fellowship," he said after a moment. He changed the topic, not wanting me to know that he was afraid.

"Yes, twice in a week", I responded.

I was finally given a schedule. The principal granted my request to host the fellowship every Wednesday for an hour. God needed a youth to rebuke him by slapping his daughter, but what the youth couldn't do was convince him not to mess with God. So, God dropped a shooting star as evidence of his presence. "And they went forth, and preached the gospel everywhere, the Lord working with them, and confirming the Word with signs followed" (Mark 16:20).

Hebrews 2:3-4: "How shall we escape if we ignore so great a salvation? This salvation, which was first announced by the Lord, was confirmed to us by those who heard him. **God also testified to it by signs, wonders and various miracles, and by gifts of the Holy Spirit** distributed according to his will."

1 Thessalonians 1:5, KJV: "For our gospel came not unto you in **word only, but also in power**, and in the Holy *Ghost..."*

My Punishment

Succeeding these events was the punishment for slapping his daughter. I was spanked and was given a week's worth of chores to do. I didn't tell anyone what I had to bear for the name of Jesus. I knew that there would be a price to pay for dedicating my life to Christ. I was willing to die for Him and I was fully persuaded that he loved me as written in the book of Romans 8:38:

"For I am persuaded, that neither death, nor life, nor

angels, nor principalities, nor powers, nor things present, nor things to come, nor height, nor depth, nor any other creature, shall be able toseparate us from the love of God, which is in Christ Jesus our Lord".

I wanted to be a true Christian, so I knew that I would have to take all kinds of accusations and beatings for God – think of Jesus's disciples. They went through sufferings for Christ, butthis was what Jesus said to them in John 15:18-20:

"If the world hates you, you know that it hated Me before it hated you. If you were of the world, the world would love its own. Yet because you are not of the world, but I chose you out of the world, therefore the world hates you. Remember the word that I said to you, 'A servant is not greater than his master.' If they persecuted Me, they will also persecute you. If they kept My word, they will keep yours also."

I believe that our enemies attack us as if they were attacking Jesus. But listen, they are liars. The battle has already been won, as it is written in John 16:33, "I have said these things to you, that in me you may have peace. In the world you will have tribulation. But take heart; I have overcome the world."

As they were praying, I was getting my spanking in the office and I showed them the mark and the swelling on my body. I told them I had to pay the price for all these people to get saved and I was like the sacrificial lamb. God didn't ask me, but I was honored to be used.

I was also like Paul who said in 2 Corinthians 11:23-29:

"I have worked much harder, been in prison more frequently, been flogged more severely, and been exposed to death again and again. Five times I received from the Jews the forty lashes minus one. Three times I was beaten with rods, once I was pelted with stones, three times I was shipwrecked, I spent a night and a day in the open sea, I have been constantly on the move. I have been in danger from rivers, in danger from bandits, in danger from my fellow Jews, in danger from Gentiles; in danger in the city, in danger in the country, in danger at sea; and in danger from false believers. I have labored and toiled and have often gone without sleep; I have known hunger and thirst and have often gone without

food; I have been cold and naked. Besides everything else, I face daily the pressure of my concern for all the churches. Who is weak, and I do not feel weak? Who is led into sin, and I do not inwardly burn?"

I had to fast. God started healing people and they were getting delivered. They began to go into the village to tell everyone. Prayer requests and invitations as a guest speaker began to stack up, new Fellowships were led off, and Churches and Fellowships began to open. These events were unplanned, but it was God's agenda to bring it to pass. I didn't know what I was doing; I was just playing by ear. I had never opened a church or started a ministry. Simply, I just picked up the Bible, read it and elected the treasurer and secretary (though technically, everybody winged it).

All I cared about was that people read their Bibles and prayed. I didn't know what to do. We didn't have any needs. We buried the money in the ground and did not spend a penny. We didn't read the people's prayer requests. We put their requests in a basket, prayed over the requests, burned them and sent them to heaven. We thought God was phenomenal. We didn't know what we were doing, but people were getting healed and delivered. The miracles were endless.

One thing that was important to me was that everybody that was saved received good grades.

Students started complaining that all of the A's were going to the students who were part of the Bible study. They started complaining that God was helping us. God helped us because we put Him first. It all lines up with the Word according to Matthew 6:33, "But seek first his kingdom and his righteousness, and all these things will be given to you as well."

Everybody was getting A's except for me. I was getting B's and sometimes less than that. I didn't understand why I had been performing less than those I teach. I decided I needed to ask the teacher why my grades were so low because I didn't want to give God a bad name. My grades were very important to me. I wanted to be a doctor, but God was leading me into full-time ministry. Colossians 3:23 says, "Whatever you do, work at it with all your heart, as working for the Lord, not for human masters". I thought of God's Word and remembered what the Lord said in Isaiah 55:8,

"For my thoughts are not your thoughts, neither are your ways my ways, saith the Lord".

My uncle who was a medical doctor and a Christian had previously asked while I was visiting his clinic and praying for the sick when I wanted to be a doctor. I told him that I like to see people get healed. He said, "Doctors cure symptoms, only God heals." Because healing was flowing through me, I considered myself to be a doctor. I stopped worrying about the grades and continued to follow the lead of the Holy Spirit. During this time, the ministry was expanding. We met every Saturday to fellowship together and at times, people would come to get me to pray for the sick and deliver those who were demon-possessed.

God is Not Pleased

Whenever we gather on the first Saturday of the month, we normally have fellowship, Bible study, words of prophecy, healing, and deliverance. We listen to every leader of every branch and we make decisions on how to handle difficult issues with the members of each fellowship. On a particular day, while we fellowshipped, during the praise and worship time, prophecy broke out – a woman spoke in tongues and interpreted what God said to us.

According to her, God said, "I will not hear you, said the Lord. You are taking my grace in vain. You are continuously living in sin. You will pay with your blood". This sounded contradictory to the Word of God, so the leaders asked me to stop the prophecy. I heeded and gave my approval to stop the person from speaking. No scripture was coming to us that lined up with what she was saying. Immediately she was stopped, the one who stopped her also started speaking in tongues and confirmed what she had spoken. It was clear God was the one speaking when two other people went to stop the prophecy and they also started speaking in tongues and prophesying the same thing.

Because I didn't understand what was going on, I closed my eyes and started to pray. I saw the eyes (red eyes) of the Lord roaming throughout our fellowship. The prophecy continued and we were instructed to fast for three days and to stop taking the Blood of Jesus in vain. This prophecy was confirmed in Romans 6:1-3, "What shall we say, then? Shall we go on sinning so that grace may increase? By no means! We are those who have died to sin; how can we live in it any longer? Or don't you know that all of us who were baptized into Christ Jesus were baptized into his death?"

My Christian life has been marred with suffering so much that it is not uncommon for me to be called out about God's desire to show me favor and love in most prophetic gatherings. I heard from the Lord about giving me planets and stars. I was puzzled and shook it off because practically nobody teaches about this phenomenon. There still exists the argument about whether a good God will destroy the earth and heaven he created. Since I became a Christian as a child and God has been faithful to everything I believed without exhaustive interpretation, I know that when God wants me to believe new revelation, He will lead me to a specific Bible verse. Meanwhile, I watch how God sets example of balance through teaching us to love our enemies by allowing us the same trial. He rewards for our belief in Him whether we pass or fail in the flesh because he contributes our failure to Himself and gives us credit for our victories.

"But I tell you, love your enemies and pray for those who persecute you, that you may be sons of your Father in heaven. He causes His sun to rise on the evil and the good, and sends rain on the righteous and the unrighteous. If you love those who love you, what reward will you get? Do not even tax collectors do the same?" (Matthew 5:44-46)

Philippians 2:15: "so that you may be blameless and pure, children of God without fault in a crooked and perverse generation, in which you shine as lights in the world."

1 Corinthians 10:13: "There hath no temptation taken you but such as is common to man: but God is faithful, who will not suffer you to be tempted above that ye are able; but will with the temptation also make a way to escape, that ye may be able to bear it."

We Christians know the way. Let us not forget that we are strangers in this land, and it is never a sad thing to go home. If we are left here, we need to continue to know that our God is faithful and will take care of every issue we have. The goal is to overcome and receive a mind-blowing reward. This is reason Jesus ran the race for the price set before him. I am choosing to help us focus on the price so we will overcome this world.

God is God, and He grows His children; however, He knows best. My expectations are sometimes not the way God chooses to speak.

My sister Silvia came to visit in 08/08/2008 and wanted to go to Larry Huck ministries. While in line to be laid hands on for impartation during the prayer session, the young minister (whom I believe is Pastor Hock's son, whether biological or not, I do not know) started to say, "Thank you for the planets, stars and universes you are giving to her." Then he felt like I grabbed his mouth because he thought he was wrong. I started to laugh and my sister wanted to know why but I decided I was not going through that rabbit hole. I said nothing. I wanted to tell him at a later time that I also heard the same thing.

On my birthday, my son, whose prophetic gift was just developing, registered planets and stars in my name. It was probably because he heard what we had been hearing and didn't know what to do with it. On a fateful day at Antioch church, I was under a ministration of a guest speaker who was prophesying. According to him, he saw the glory light on me, prayed for me and pleaded with God to make the later part of my ministry easier. Likewise, the Assemblies of God church I typically attended also saw the glory light and wanted to pray with me. You may want to know what this glory light is.

1 Peter 4:14: "If you are insulted because of the name of Christ, you are blessed, for the Spirit of glory and of God rests on you."

Fools for Christ

1. **1 Corinthians 4:10-14:** We are fools for Christ, but you are wise in Christ. We are weak, but you are strong. You are honored, but we are dishonored! To this very hour we are hungry and thirsty, we are poorly clothed, we are brutally treated, we are homeless. We work hard with our own

hands. When we are vilified, we bless; when we are persecuted, we endure it. when we are slandered, we answer gently. Up to this moment we have become the scum of the earth, the refuse of the world. I am not writing this to shame you, but to warn you as my beloved children.

Even when we are reviled here, our blessings gather here and in the life after.

1 Peter 4:14 says the GLORY and the Spirit of God rests on you. The spirit of God is the same one that descended on Christ at baptism and said, “This is my beloved son in whom I AM PLEASED”. How much will you pay to be loved by God the same way He loved Jesus Christ? This type of confidence comes through trials, when you prove to God that He is everything and always first in your life.

In our society today, there are many shifts happening – unexpected trials and heartbreaks. I pray that the spirit of peace and comfort rests on everyone who needs it. I truly pray for an encounter or surrender of souls by their return to God. As no one can control the wind, no one has control of their lives. So, the pandemic we are currently in can only be controlled by God. While many families have been devastated beyond words, my concern is about the living. This type of sickness, or one similar to it, can still reoccur without warning, just like this one, and we may not be ready. If you are not in a community like church, Bible study or fellowship, out of love, I beseech you to do so.

Some people have been asking if this is a sign of the end. I tell you that the end started the moment you were born. The Bible said the moment Jesus was born, the wisemen saw his star moving.

Matthew 2:2: "Where is the one who has been born king of the Jews? We saw his star when it rose and have come to worship him."

Matthew 2:9: “After they had heard the king, they went on their way, and the star they had seen when it rose went ahead of them until it stopped over the place where the child was.”

They did not say they saw his star born or created, rather they said they saw it rise. The Bible says you have been predestined for good work, you have been created for his service and he equips you with every tool you need to perform his assignment.

God Works in All Things

1. Romans 8:29-31: "For those God foreknew, He also predestined to be conformed to the image of His Son, so that He would be the firstborn among many brothers. And those He predestined He also called; those He called He also justified; those He justified He also glorified. What then shall we say in response to these things? If God is for us, who can be against us?"
2. **Hebrews 9:27**: "It is appointed unto man once to die and after that it is judgement. And as it is appointed unto men once to die, but after this the judgment. Yet there are many accidental deaths every year. God is not unaware."
3. **Psalm 118:6:** "The LORD is with me; I will not be afraid. What can mere mortals do to me?"

Chapter 3

GOD'S EYE

Julie's Vision

Many people were speaking about what God was saying. I raised my hand and I told them what I had seen in a vision: God's eyes roaming the fellowship hall. Afterward, I spoke to the people and asked them to obey the word of the Lord: to fast and seek God's face for three days. I told them we would follow this up by gathering to worship the Lord, seek His face and then He would hear us. The people all went back to their villages and dormitories to commence the fasting.

Among us was a student, Patience, who had just joined us at the fellowship and was being persecuted by her father. Her dad decided to withhold her school fees because of her conversion to Christianity. I took this personally by borrowing money from my mother's driver to pay her fees. I hoped to refund the money before my mother returned from London. I was weeping before the Lord and during my fellowship with him. I tried to resign from my leadership role. I was saddened because God was displeased with us. I felt badly about it.

Later on, I was put in a trance by the spirit of the Lord and Jesus appeared to me. I was apologizing to him for my failures and I told him that I wanted to resign and just go to heaven, instead of trying to lead this mission and fail.

Patience's Vision

As I was praying, Patience ran into the room, knelt down and she began to repent from her sins. I fell into a trance; I felt the weight of my sins. An angel of the Lord took the weight from her and the other Angel wrote her name in the Book of Life. The angels reminded her that it was her birthday – her spiritual birthday to be precise.

During my vision, I asked Jesus to take me with him because I didn't want to go to hell.

Jesus pointed and said, "Over there, God is".

As I looked up, I saw Father Abraham, Isaac and Jacob and the sons that came before them.

"Do I know these people?", I thought. I didn't know Father Abraham, Isaac and Jacob at the time. I'd never seen them. But this was how they introduced themselves. They told me that Sister Julie was in the corner crying, "Go and tell her that we are rooting for her."

I asked Patience, "They called you by name?".

I shut up and I listened. Then I thought, *Well, I guess I am not as bad as I thought I was. If Father Abraham calls me by my name, then that is where I'm going when I die. Wow! They are watching me and they trust me. They confirmed that I didn't screw up earlier. Amazing.* I wiped the tears off my face and I looked back at the wall to see what Jesus' opinion was. He came there to confirm that he sent her. He disappeared after this revelation.

From that day onward, I have been feeling like this was the greatest thing that has happened to me as a Christian. I have been trying to be holy and live a righteous life. Now, all eyes are watching me. The saints that have come before and all of heaven watched me. This lined up with the Word of God according to Hebrews 12:1, which says, "Wherefore seeing we also are compassed about with so great a cloud of witnesses, let us lay aside every weight and the sin which doth so easily beset us, and let us run with patience the race that is set before us."

They called me by name because they knew I was about to quit. I hadn't told anyone that I wanted to quit. It's good to know that these people know you by name. That's awesome. Some people don't believe in the Bible. But my middle name is the Word of God.

Apparently, God expects me to live it out and heaven is helping me. With serious prayer, I will not let the heavenly team down.

We continued to fast and pray in the middle of our school campus where we had planted a garden. Gold dust started descending from heaven. Gold was all over our hands and bodies. We were thrilled. I thought God was helping us. We were seeing visions and our lives were being transformed. God is good. God is faithful. He forgave us! We are reunited with Him; everything was going fine. Our fellowship continued.

Julie Raises the Dead

People started taking me to their dead in order to pray for them. The first person that I prayed for didn't get up because I hadn't obeyed God's instructions. I didn't touch him and I didn't want to touch him. He was dead and I was just a child. It had been a tough journey. I was so frustrated. The dead person was eventually buried.

On my way from the funeral, God sent another brother to take me to a brother who had been dead for 4 days. Someone had a vision that He would come back to life. Mind you, I was a visitor in the area and didn't know anyone. I was sent to school by my parents who do not share my faith. The first person I prayed for did not get up. Now men in a motorcycle (a vehicle I hate because I have watched a lot of people die by falling off from it) wanted me to go to a place and pray for a man who has been dead for four days on a motorcycle. I was really disappointed that the pregnant woman I just prayed for did not rise. I was worried about the faith of the new converts with me, but I was even more afraid of the whole thing. I knew God could raise the dead but regarding the pregnant woman, God said I should touch her. I was still garnering strength, but the family decided to go and bury her. Sad and confused, too young to negotiate, I blamed God.

Reluctantly, I got on the motorcycle to where the dead man was placed. After saying some prayers, he coughed a smoke-like substance out of his mouth and came back to life. He woke up and said, "Amen". I did not know what to do afterward. I wanted to run away because I have never seen a dead man rise. I simply believe that a God who made us can fix us. The Bible told us to raise the dead and do more than Christ did so men can come to God. I said I needed to go but the people assumed I had a full schedule. They said there were other dead individuals to pray for in the villages. I had to let them know that I am not that special; I am only a child. I didn't know why God chose to use me, but I was thankful. People stood in awe of what that just happened. Absolutely, it is a wonder when you see the Word come alive.

It Has Been a Journey

God was performing so many miracles that I couldn't keep count. Right after that, I started moving from university to university, fasting, praying, preaching, and watching God save lives. I preached about the rapture and it got every student on their toes, waiting for the signs.

One day, while I was in school, I asked my math teacher if I could study with her. She said, "Yes" and took me to her place. While I was there, I had a vision of fine china descending from heaven. I was in awe and began to thank the Lord. I praised him.

The Rain and the Crusade

At the same time, it began to rain. This prompted several students from the school to go into the city for their groceries and there, they were greeted by angels who said, "Julie is looking for you". They were frightened. When they got back to the campus, they couldn't find any of the leaders of our fellowship. It seemed that God had hidden us all. They began to weep because they thought that rapture had happened; they thought God had left them. They began to run around the campus looking for me. About two hundred students searched the campus until they found me in the teacher's apartment.

"Oh Julie, you are here! Did you know that the rapture happened and everybody else is gone," they said.

"No," I said. "The rapture has not happened". I told them that the rapture had not occurred because according to Revelations 1:7, every eye will see him. And here is the scriptural reference: "Look, he is coming with the clouds," and "every eye will see him, even those who pierced him"; and all people on earth "will mourn because of him." So shall it be! Amen (Revelations 1:7).

I began to ask them, "Are you ready?" I stated that God is teaching you to be ready for Christ's return twenty-four hours a day.

I taught them Jesus may not come in their lifetime, but that they will surely meet him. I said that people die every minute of the day and that you are scared because you are not ready. I started teaching them that our God is a consuming fire and they have heard the warning according to Hebrews 12: 22-29 which says,

"But ye are come unto Mount Sion, and unto the city of the living God, the heavenly Jerusalem, and to an innumerable company of angels, To the general assembly and church of the firstborn, which are written in heaven, and to God the Judge of all, and to the spirits of just men made perfect, And to Jesus the mediator of the new covenant, and to the blood of sprinkling, that speaketh better things than that of Abel. See that ye refuse not him that speaketh. For if they escaped not who refused him that spake on earth, much more shall not we escape if we turn away from him that speaketh from heaven Whose voice then shook the earth: but now he hath promised, saying, yet once more I shake not the earth only, but also heaven. And this word, yet once more, signifieth the removing of those things that are shaken, as of things that are made, that those things which cannot be shaken may remain. Wherefore we receiving a kingdom which cannot be moved, let us have grace, whereby we may serve God acceptably with reverence and Godly fear: For our God is a consuming fire".

I looked for the sisters to write down the names of those that responded to the altar call, but they weren't available. Lo and behold, they walked in during the altar call, came to me and said that they were talking to a lady who said that Julie was looking for them. Later, they realized that it was an angel. God was gathering the people. God was staging a crusade. I continued preaching and telling them about Jesus. God made me stay indoors; He's all-knowing. He knew where and when the students would gather. God sent an angel beforehand to gather the officers of the ministry who would follow up those who would give their lives to Jesus. This is another example of predestination.

People wanted to be saved, but they didn't know how. We were teenagers praying for revival. We were willing to show them how to be saved, but we didn't know how to plan a crusade. God is ultimately in charge. He answered our prayers by bringing a fierce storm.

The Gift of God

[28] From the standpoint of the gospel, they are enemies for your sake, but from the standpoint of God's choice, they are beloved for the sake of the fathers; [29] for the gifts and the calling of God are irrevocable. [30] For just as you once were disobedient to God, but now have been shown mercy because of their disobedience, [31] so these also now have been disobedient, that because of the mercy shown to you they also may now be shown mercy. [32] For God has shut up all in disobedience so that He may show mercy to all.

(Romans 11:28-32)

God's gift is irrevocable. God was pleased with the Word that came out of my mouth which could have brought healing to God's people. I had to learn how to put my flesh aside. As it is written in Galatians 5:16-18, "But I say, walk by the Spirit, and do not gratify the desires of the flesh. For the desires of the flesh are against the Spirit, and the desires of the Spirit are against the flesh; for these are opposed to each other, to prevent you from doing what you would. But if you are led by the Spirit you are not under the law".

Fear God Alone

1. **Matthew 10:27-29**: 27What I tell you in the dark, speak in the daylight; what is whispered in your ear, proclaim from the housetops. 28Do not be afraid of those who kill the body but cannot kill the soul. Instead, fear the One who can destroy both soul and body in hell. 29Are not two sparrows sold for a penny? Yet not one of them will fall to the ground apart from the will of your Father.

According to CNBC TV, WORLD NEWS (29th Aug. 2018), Worldwide gun deaths reached 250,000 yearly; US ranks high. The following 2017 data (for USA) have been retrieved from The Associated Press, National Center for Health Statistics, Deaths and Mortality Section:

- Number of deaths: 2,813,503
- Death rate: 863.8 deaths per 100,000 population

- Life expectancy: 78.6 years
- Infant Mortality rate: 5.79 deaths per 1,000 live births

2. **Luke 12:5:** But I will show you whom you should fear: Fear the One who, after you have been killed, has power to throw you into hell. Yes, I tell you, fear Him!

3. **Hebrews 10:31**: It is a fearful thing to fall into the hands of the living God.

4. **2 Samuel 12: 1- 25:** 1The Lord sent Nathan to David. When he came to David, he said, "There were two men in a city. One was rich, but the other was poor. 2The rich man had many sheep and cattle. 3But the poor man had nothing except one little female lamb he had bought. The poor man fed the lamb, and it grew up with him and his children. It shared his food and drank from his cup and slept in his arms. The lamb was like a daughter to him.

[4]"Then a traveler stopped to visit the rich man. The rich man wanted to feed the traveler, but he didn't want to take one of his own sheep or cattle. Instead, he took the lamb from the poor man and cooked it for his visitor."

[5] David became very angry at the rich man. He said to Nathan, "As surely as the Lord lives, the man who did this should die! [6] He must pay for the lamb four times for doing such a thing. He had no mercy!"

[7] Then Nathan said to David, "You are the man! This is what the Lord, the God of Israel, says: 'I appointed you king of Israel and saved you from Saul. [8] I gave you his kingdom and his wives. And I made you king of Israel and Judah. And if that had not been enough, I would have given you even more. [9] So why did you ignore the Lord's command? Why did you do what he says is wrong? You killed Uriah the Hittite with the sword of the Ammonites and took his wife to be your wife! [10] Now there will always be people in your family who will die by a sword, because you did not respect me; you took the wife of Uriah the Hittite for yourself!'

[11]"This is what the Lord says: 'I am bringing trouble to you from your own family. While you watch, I will take your wives from you and give them to someone who is very close to you. He will have sexual relations with your wives, and everyone will know

it. 12 You had sexual relations with Bathsheba in secret, but I will do this so all the people of Israel can see it.'"

13 Then David said to Nathan, "I have sinned against the Lord."

Nathan answered, "The Lord has taken away your sin. You will not die. 14 But what you did caused the Lord's enemies to lose all respect for him. For this reason, the son who was born to you will die."

15 Then Nathan went home. And the Lord caused the son of David and Bathsheba, Uriah's widow, to be very sick. 16 David prayed to God for the baby. David fasted and went into his house and stayed there, lying on the ground all night. 17 The elders of David's family came to him and tried to pull him up from the ground, but he refused to get up or to eat food with them.

18 On the seventh day the baby died. David's servants were afraid to tell him that the baby was dead. They said, "Look, we tried to talk to David while the baby was alive, but he refused to listen to us. If we tell him the baby is dead, he may do something awful."

19 When David saw his servants whispering, he knew that the baby was dead. So he asked them, "Is the baby dead?"

They answered, "Yes, he is dead."

20 Then David got up from the floor, washed himself, put lotions on, and changed his clothes. Then he went into the Lord's house to worship. After that, he went home and asked for something to eat. His servants gave him some food, and he ate.

21 David's servants said to him, "Why are you doing this? When the baby was still alive, you fasted and you cried. Now that the baby is dead, you get up and eat food."

22 David said, "While the baby was still alive, I fasted, and I cried. I thought, 'Who knows? Maybe the Lord will feel sorry for me and let the baby live.' 23 But now that the baby is dead, why should I fast? I can't bring him back to life. Someday I will go to him, but he cannot come back to me."

24 Then David comforted Bathsheba his wife. He slept with her and had sexual relations with her. She became pregnant again

and had another son, whom David named Solomon. The Lord loved Solomon. [25] The Lord sent word through Nathan the prophet to name the baby Jedidiah, because the Lord loved the child".

There are many people who misunderstand God's righteousness. God is not looking to punishing people for what they have done or not done. The Lord wants us to obey him and trust him.

Carry One Another's Burdens

Galatians 6:6–8: [6]But the one who receives instruction in the word must share in all good things with his instructor. [7]Do not be deceived: God cannot be mocked. Whatever a man sows, he will reap in return. [8]The one who sows to please his flesh, from the flesh will reap destruction; but the one who sows to please the Spirit, from the Spirit will reap eternal life....

Job 4:8: As I have observed, those who plow iniquity, and those who sow trouble, reap the same.

Job 13:9: Would it be well when He examined you? Could you deceive Him like a man?

The man after God's heart is God's endearing term to David. This is what he saw in the star of David. Lo and behold, sin entered his heart and tried to destroy him. Remember Abraham, the friend of God" He had a child after the flesh, Ishmael, and God did not want that for David. God does not want to look at David and have a constant reminder of when David did not come to Him with his faults. He had to be confronted before He went to God.

Acts 13:22: After removing Saul, he made David their king. God testified concerning him: 'I have found David son of Jesse, a man after my own heart; he will do everything I want him to do.'

Although God allowed his child to die, God gave him Solomon, the richest, wisest man that ever lived. Although God allowed Abraham's son to live, Abraham's heart till date bears the sadness of watching his children fight. God does not even consider Ishmael as Abraham's son (Genesis 15:4 says, "Then the word of the LORD came to him: 'This man will not be your heir, but a son who is your own flesh and blood will be your heir.'")

We claim children because of physical relations. God sees all children as his. The angel of the Lord heard Ishmael's cry and blessed him (Genesis 21:17- 18 says, "[17]God heard the boy crying, and the angel of God called to Hagar from heaven and said to her, "What is the matter, Hagar? Do not be afraid; God has heard the boy crying as he lies there. [18]Arise, lift up the boy and take him by the hand, for I will make him into a great nation.")

The star of Ishmael will be a great nation. Today, lots of Muslims are having dreams of Christ, giving their life to Christ and are being ingrafted in Abraham. God cleans up our messes when he knows we are not capable of cleaning them up. He is a good father.

Therefore, if our situation is caused by sin or an inadvertence, be assured that God will extend the same love and care as he did for those before us. We must give him the permission to do so by going to him.

2 Samuel 12:13 says, "Then David said to Nathan, 'I have sinned against the Lord.' Nathan replied, 'The Lord has taken away your sin; you will not die'". When we confess our sins, God is faithful to forgive us. Christ has died for our sins from the foundation of the world and all we need to do is ask for forgiveness.

Ambassadors for Christ

1. **2 Corinthians 5:20-21**: 20Therefore we are ambassadors for Christ, as though God were making His appeal through us. We implore you on behalf of Christ: Be reconciled to God. 21God made Him who knew no sin to be sin on our behalf, so that in Him we might become the righteousness of God.

I found the pictures used in this book online and it is fitting in illustrating how God sees us and the choices we make to listen and follow Christ by accepting his sacrifice and letting his spirit guide us.

Chapter 4

THE LORD SPEAKS ABOUT JULIE'S BOOK

I heard a voice say, "Julie Ann".

While singing, I kept walking towards the shower and I heard the voice again, "Julie Ann. Your story will be read as one of the stories of the Bible. Your story will be read like the Bible."

"*Wow!*" was my first response. I knew that God was going to create a new heaven after this, but I have known something else that nobody else knew. I always thought to myself that after this heaven and earth, there was going to be another Bible and that my story would be included. That was my understanding of the message. I thought maybe my story would be written like the story of Mary, Martha, or the Good Samaritan. I was so happy, and my mouth was full of thanks to God. Then He told me, "Julie, start writing the book".

If I had known that God wanted me to write a book, I would have documented my story. Since the host of Angels and Abraham watched me, I assumed that the book would be written in heaven. I like the way Eric Lyons with the Apologetics Press explained it in the "Image and Likeness of God". These are men of God filled with love and grace of our Lord Jesus Christ. I concurred with many of their assertions but for the purpose of this book, I will tell you what the Lord showed me about Himself in 2007 in addition to what I already told you about an experience I had during my teenage days.

I came back from Bible Fellowship, I was at the balcony walking away from my mother who objected to my Scripture Union

Bible study relationships because I went to the fellowship without her knowledge. I was relieved, praising God in my spirit as I galloped into my room. I had already heard the voice of the Lord while in the shower, but I dismissed it. During a school break, shortly after I had been told that father Abraham, Isaac and Jacob and the host of angels were rooting for me to make it, I heard an audible voice that said your story will be told around the world. Ever since then, my outlook about the sky changed. I was innately aware that every action I take is documented and watched closely. I believe I paid more attention this time because I was revived at the fellowship so that the flesh was subdued enough for God to have more access to me.

Hebrews 12:1-3: Wherefore seeing we also are compassed about with so great a cloud of witnesses, let us lay aside every weight, and the sin which doth so easily beset us, and let us run with patience the race that is set before us, looking unto Jesus the author and finisher of our faith; who for the joy that was set before him endured the cross, despising the shame, and is set down at the right hand of the throne of God. For consider him that endured such contradiction of sinners against himself, lest ye be wearied and faint in your minds.

At the time, I thought that either in the new heaven or the new earth, my persecution and sufferings would be mentioned in the new Bible and would be used in those days as a way to encourage the next generation who will run the race. Imagine how surprised I was when the Lord in 2007 said "Now write the book".

"I thought you were writing the book", I said.

I have not kept dates, times or any material that will be proof that the events I am to share are true. In my opinion, this is the way He likes it. God is timeless – everlasting. He is the same yesterday, today and forever and He never changes. He is consistent with his style that dates and styles are not important for this book.

In 2007, I was going through a very trying time that I almost made me not want to continue the ministry. I was not strong enough to handle the associating pressor of living a holy life. I did not want to backslide so I prayed not to wake up every day. "Take me home, take me now, God." I prayed for six months. Instead, He continued telling me to write the book. Although I did not know if I could make it or not, God's answer was, "Not now; the book is yet to be completed".

Shortly after, I was deadly sick and this experience made me believe. I was so sick that I had to crawl from the bedroom to the bathroom just to ease myself. I was unable to go to the companies I owned or manage my rental properties and my son. I was resolved to going to heaven, so I continued to wait for God to take me home.

I have really struggled with whether or not I should tell my story. My biggest fear always was, "Will anyone believe me?" I mean, I found it even harder to believe myself! As I began to talk about it, I got accused of very bad things – some that are not godly. It is amazing that the biggest opposition I got was from my fellow believers – those who were supposed to encourage me. Most people think it is hyperbole when I tell them that the Lord took me to heaven when I was about 11 years old. Some openly told me that my exaggerated imaginations were being fueled by some unclean spirits.

But mine has been an interesting journey. When the Lord first talked to me with an audible voice when I was only nine, I didn't know what to think of it. In fact, I instinctively thought the devil was trying to trick me by masquerading as an angel of light. So, in my ignorance, my first reaction was to avoid intimacy with God. I didn't want to hear any more voices or see more angels. I asked God to show me nothing unless He used His Word to reveal it to me. I let him know I was afraid of being misled so I wanted to be sure it was God. God was gracious and He continued revealing himself in His Word as well as through visions and the audible voice. He also started using me to do diverse miracles including deliverance of those possessed by the devil as well as raising the dead.

My story is so fascinating that it is hard to compile in one book but I have decided to take it one step at a time. The thought of writing a book had not crossed my mind until the Lord spoke to me in 2007 and instructed me to document my experience in a book. Since this was the last thing on my mind, I hadn't been preparing by getting some chronological data compiled. So, I didn't know how to go about it. But the Lord drew my attention to how the Bible was written - authors were inspired by God's spirit and they knew what to write. And that's the approach I took. I am totally depending on the Holy Spirit to guide me on what to pen down.

As soon as I got the instruction to write the book, I went through a series of attacks that almost made me give up. I kept asking God to take me home. But God had better plans for me – instead of taking me home, he brought home to me. I recalled attending Benny Hinn's Conference in Gaylord Texas. I had felt the Holy Spirit leading me to attend the event, so I attended the conference with my son. At this conference, I met Jimmy Laurence, a gospel minister. He told me the Lord wanted me to visit him. I dismissed the invite because I didn't know him too well, so I didn't know what to make of it.

This was an interesting time for me – even though some of my businesses were still functioning, I was constantly battling with bad health. Since I couldn't be there physically, my employees would call me every day so we could pray together and dedicate the day to God. This also helped me to be in the loop.

One day, I was talking to Trena, my massage therapist while lying in bed. Suddenly, a hand appeared over my bed. I was so perplexed that I shouted to Trena that there was a hand in my room. She advised me to call the police, but I knew the police wouldn't know how to help me.

While we were still talking, I saw an incoming call and I picked it up. It was Silvia Miranda, my son's godmother. She was calling from Denver, where she lives. Silvia is like a family member and I am always happy to talk with her. She is usually a bit reserved but this time, I could tell she was excited over the phone.

"Julie! You will not believe what just happened!" she yelled. But I didn't want to know. All I was thinking about was the hand in the room.

"Silvia, there is a hand in my room!" I shouted back at her. She either didn't hear me or considered what I was telling her not as exciting as what she wanted to share.

"Julie, as I was praying for you, I saw an opened heaven and you were at the feet of the Lord," she said.

Big deal! There was a hand in my room! That was more urgent and more important. So, I repeated for the second time, "There is a hand in my room!" As I said that, it suddenly dawned on me that God was trying to get my attention. I recalled what Ms. Laurence had tried to tell me earlier - that God was trying to get my attention. Right there, I made up my mind to give God the attention He was trying to get from me.

I wasn't sure if I was in trouble for my unbelief and I was unsure how God would deal with me. I immediately told Silvia that if she didn't hear from me again, she should know that I was either dead or God had taken me to heaven. I also asked her to come and pick up my son. As soon as I hung up, I called the school to find out if my son was okay. The school confirmed he was doing fine, so I turned to prayer, asking God for forgiveness.

As I was praying, my entire house was transformed, and I could see the sky from my room. I could see the 4th heaven, where the Lord was seated. This kept happening regularly for 3 and a half years. When I first narrated my first vision of heaven, most people thought I was just imagining things. But this is more than just a figment of my imaginations. It is my prayer that God will reveal the truth to people because the Bible says God has the power to turn the hearts of men wherever he wills.

"The king's heart is in the hand of the LORD, as the rivers of water: he turneth it whithersoever he will." Proverbs 21:1 (KJV)

Of all people, I don't know why God chose to reveal some revelations to me. But I am a willing vessel in His hands. And I will do my part - I will tell the world what He told me to share with them. I will share the secrets He revealed to me when He took me to heaven on various occasions – and I hope this will help to convince some skeptics that there is a living God.

"Wow! God is going to create a new heaven after this; I know something nobody knows. After this heaven and earth, there is going to be another Bible that my story will be written into", I soliloquized.

That was my understanding of the message. I was so happy. I thanked God. Furthermore,

God told me, "Julie start writing the book". Since I know what it costs to disobey him, I did the noble thing – I obeyed and started writing and here we are! I cannot begin to write without remembering.

The visit of Father Abraham, Isaac and Jacob to Sister Patience to bring me the word of hope ensured the continuity of the ministry. Abraham was called a friend of God and also the father of all Nations. God made a covenant with Abram in Genesis 15:5-6:

5 *And the LORD took him outside and said, "Now look to the heavens and count the stars, if you are able." Then He declared, "So shall your offspring be."* 6 *Abram believed the LORD, and it was credited to him as righteousness.*

Chapter 5

WHY STARS?

Genesis 1:16: *God made two great lights-the greater light to govern the day and the lesser light to govern the night. He also made* ***the stars****.*

Psalm 147:4: *He determines the number of* **the stars** *and calls them each by name.*

I guess that God told him my name since He knows the names of every star. The star fell when God used me to start a ministry and after the believers' judgment on the earth, God will give a star to those that overcame and helped him with the leadership of the new heaven.

I have been a minister for over 40 years and have heard lots of sermons about heaven. However, I have never heard anyone speak correctly about our rewards when we get to heaven**.** Everyone agrees we will go to paradise and the Bible tells us of a new heaven.

A New Heaven and a New Earth

The idea of a new heaven and a new earth is explicitly noted in Isaiah 65:17; 66:22; 3:13; and Rev. 21:1. The Old Testament has no term that directly translates as "universe"; the phrase "heaven and earth" was the Hebrew way of referring to the universe they knew and then the Bible made an elaborate description in Revelation 21 about the capital city of heaven. Store your treasures in heaven, where **moths and rust cannot destroy**, and thieves do not break in and steal. According to the English Standard Version, it said, "but lay up for yourselves treasures in heaven, where neither moth nor rust destroys and where thieves do not break in and steal" (Matthew 6:20).

In heaven, the work does not stop, as sitting on the throne means ruling or leadership. There is no room for idleness. In Rev. 3:21, God said to the Churches in Laodicea:

"20Behold, I stand at the door and knock. If anyone hears My voice and opens the door, I will come in and dine with him, and he with Me. 21 To the one who is victorious, I will grant the right to sit with Me on My throne, just as I overcame and sat down with My Father on His throne. 22He who has an ear, let him hear what the Spirit says to the churches."

Jesus is the first fruit, and all believers are the rest of the fruit of God. Jesus is not just sitting in heaven; He is interceding for us.

Romans 8:34: God Works in All Things

33Who will bring any charge against God's elect? It is God who justifies. 34Who is there to condemn us? For Christ Jesus, who died, and more than that was raised to life, is at the right hand of God—and He is interceding for us. 35Who shall separate us from the love of Christ? Shall trouble or distress or persecution or famine or nakedness or danger or sword?

He is walking as God among us; He is visiting us.

1 Corinthians 15:8-9: 8And last of all He appeared to me also, as to one of untimely birth. 9For I am the least of the apostles and am unworthy to be called an apostle, because I persecuted the church of God.

It's a privilege to receive nations and stars or to have a universe that has planets. That may be why He says no eye has seen what He has prepared for those that love him.

Revelation 2:28: 28I will also give that one the morning star.

What do we know about stars?

Genesis 17:1-2, 7-9: "And when Abram was ninety years old and nine, the LORD (Yahweh) appeared to Abram, and said unto him, I Am the Almighty God; walk before Me, and be thou perfect. And I will make My Covenant between Me and thee, and will multiply thee exceedingly... And I will establish My Covenant between Me and thee and thy seed after thee in their generations for an Everlasting Covenant, to be a God unto thee, and to thy seed after thee. And I will give unto thee, and to thy seed after thee, the land wherein thou art a stranger, all the land of Canaan, for an everlasting possession; and I will be their God. And God said unto Abraham, thou shalt keep my covenant therefore, thou, and thy seed after thee in their generations."

In the mid-1990s, astronomers found strong evidence of planets around other stars.

Disclaimer: This material is being kept online for historical purposes. Though accurate at the time of publication, it is no longer being updated. The page may contain broken links or outdated information, and parts may not function in current web browsers.

How many solar systems are in the galaxy?

I have a question from our young friends at the Mountain Home Air Force Base Youth Activities Center in Mountain Home, Idaho. They wonder how many solar systems are in the galaxy. Well, I wish I knew the answer to this thought-provoking question, but not only do I not know, no one does. Even so, this question brings up some thrilling ideas.

For many years, scientists have studied the solar system. But until the last few years, we knew of no other solar systems. This may seem surprising, as the Sun is one of about 200 billion stars (or perhaps more) just in the Milky Way galaxy alone. With all those other stars, why haven't scientists studied other solar systems, at least enough to know how many are in the galaxy?

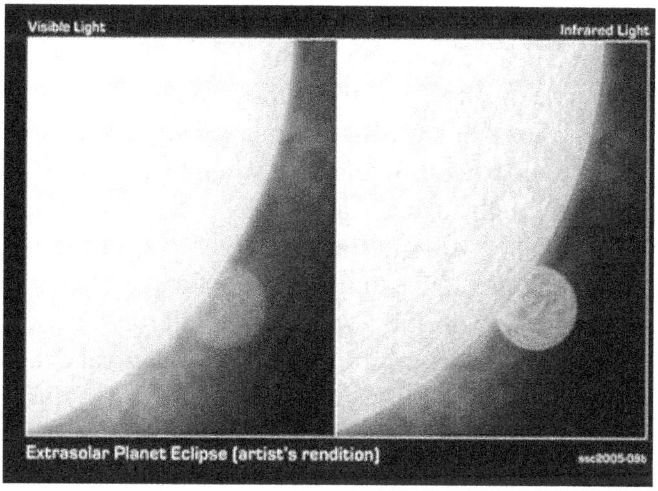

Well, the reason is that planets around other stars are really hard to find. Using regular visible light telescopes, planets are very hard to see in the glare of a star. Using infrared space telescopes, the planets show up much more clearly. Planets shine only by the light they reflect from the star they orbit, and they still don't reflect much light at that.

The stars, along with any planets under their control, are so far away that picking out a faint planet near a distant star is like spotting a mosquito next to a brilliant searchlight some miles away.

Marc Rayman, at age 14, meets astronomer Peter van de Kamp (center), who had "discovered" planets outside our solar system. On the right is radio astronomer Grote Reber. (Image from Sky and Telescope, Aug. 1971)

Although scientists, philosophers, writers, and people like you who have been fascinated by the universe have thought about other solar systems for centuries, they haven't had any to study. When I was young, this was one of many topics that I spent a great deal of time wonderingabout. In fact, when I was in the ninth grade, I was lucky enough to meet an astronomer who thought he had detected two planets around Barnard's Star, one of the closest stars to our solar system. It was quite a thrill for me to meet someone involved in such an exciting work. Alas, later evidence suggested his conclusions were incorrect, but I learned a great deal about the subject, as well as about the scientific method, by studying what this impressive astronomer had accomplished.

Finally, in the mid of 1990s, astronomers found strong evidence of planets around other stars. In all cases, they didn't find the planets by taking pictures of them, but rather by detecting their astonishingly gentle tugs on the stars they orbit. Although the star holds the planet tightly in its gravitational grip, the planet also exerts a gravitational pull back on the star, and that is what astronomers measure. It amounts to seeing the star wobble back and forth very slightly as the planet completes each orbit. Learn more about this gravitational dance as you try to solve the extraterrestrial riddle.

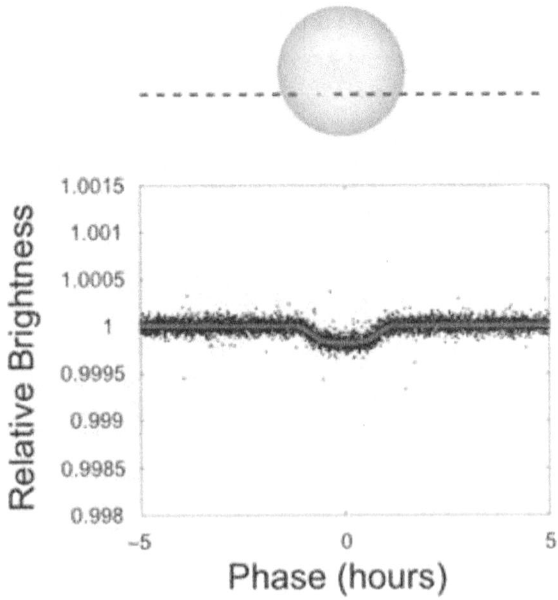

Brightness of star dips as planets passes between it and Earth

Later in time, astronomers started detecting planets through several other methods as well. For example, if the orbit of a planet happens to be aligned so that planet occasionally travels in front of the star from our perspective on Earth, it blocks some of the light. Even though the planet is tiny compared to the star, extremely sensitive instruments can measure the tiny change in brightness. NASA's Kepler mission used this technique to identify hundreds of stars that may have planets. Astronomers are observing these stars more carefully to confirm the presence of the candidate planets.

NASA is working on more space missions that will allow scientists not only to find other solar systems but also to study the planets there in greater detail. Some of the intriguing questions these missions might help answer are how common are other solar systems; is our solar system typical, with giant planets like Jupiter and smaller ones like Earth; how do solar systems form and evolve; are there other planets capable of supporting life, and is there life on other planets?

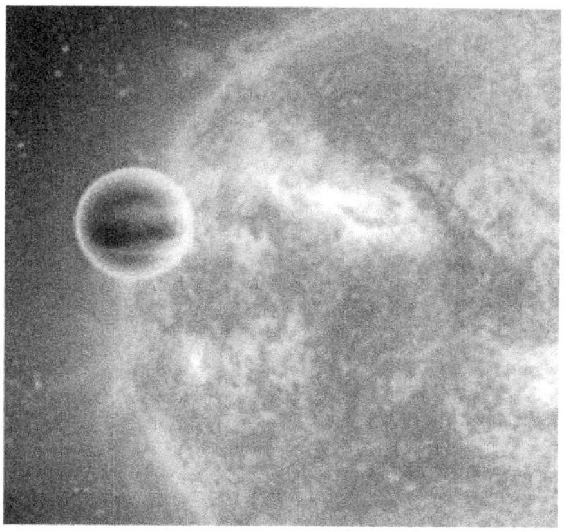

Artwork shows a steaming hot (with water!) planet discovered in another solar system.

So far, astronomers have found more than 500 solar systems and are discovering new ones every year. Given how many they have found in our own neighborhood of the Milky Way galaxy, scientists estimate that there may be tens of billions of solar systems in our galaxy, perhaps even as many as 100 billion.

Whether this estimate is correct and how similar other solar systems are to ours, remainsto be seen. It has only been a few years since the first solar system was detected apart from ours, and they are still extremely difficult to study, so this whole subject is still in its infancy. By the time our friends who asked the questions are adults, we will know a great deal more. Perhaps someday, you will help find the answers. And even if you don't, you may grow up in a time when humankind has a much clearer idea of how we and our home planet fit into the cosmos.

Psalm 19:1: The heavens declare the glory of God; the skies proclaim the work of his hands. Day after day they pour forth speech; night after night they display knowledge.

Isaiah 65:17: For, behold, I create a new heaven and a new earth: and the former shall not be remembered, nor come into mind.

Proverbs 25:2: The honor of God is to hide a thing, and the honor of kings to search out a matter.

2 Peter 3:13: But in keeping with his promise we are looking forward to a new heaven and a new earth, where righteousness dwells.

Numbers 24:17: "I see him, but not now; I behold him, but not near. A star will come out of Jacob; a scepter will rise out of Israel. He will crush the foreheads of Moab, the skulls of all the people of Sheth."

Chapter 6

FAITH VACCINATION

If I have learned anything on these adventures with the Lord, it would be the incredible importance for man to work by faith. It is only this currency that can allow you to succeed in the understanding of a God who did not only create this universe but the stars within them.

When you travel to another country whether it is your birth country or not, you vaccinate so you can be protected from unfamiliar bugs. That is not different from heaven. The only currency in heaven is faith in God and his son.

John 6:29: Jesus answered, "The work of God is this: to believe in the one he has sent."

I believe in God and I have been vaccinated for heaven because I have received Jesus Christ and made him Lord and king of my life. I have been persecuted and proved my life by becoming steadfastly unmovable in my belief of his word unto death.

When the Father called me to see him, I was terrified because I had heard of His wrath. It wasn't until I saw Him in heaven did I begin to desire an intimate knowledge of Him. It would be sad for some people to love me as much as I did and end up breaking His heart and rejecting Him by accusing Him of hate. Hate is not in Him because He is love. He knows all of us intimately and that may contribute to His interest in choosing me to manifest Himself as He did of the prophets of old.

Does Not Care About Sin

When you give birth to children, do you have rules and regulations for them to follow? Or you would watch them to see what they will do and correct them if they have an incorrect view of a thing or a situation? Your guidance comes handy to make them better at each stage of their development. See, God wanted a family with people that look and act like Him and that is why He made man.

God created mankind in his own image. But then, man ate poison (the forbidden fruit, Genesis 3:6) and it killed the soul of man (Ezekiel 18:4). It became a serious virus that remained in the human race. God wanted to help the man recover from this deadly virus. Just like when a child gets sick, whether by accident or not, their parent will try to fix it. God wanted to help us in identifying what is good from what is bad and get us to choose the good, but we demanded a law and another king other than him. As a parent, making our children happy is God's joy. He allowed us to have our way.

God would have preferred we lived without the law in the Garden of Eden but when we messed up, there was no alternative. The reason God did not want to give us the rules is because He is our maker, and he knows that based on which DNA strings we have, some rules will be hard to keep and others easy. Instead of us worrying about the rules, we should just serve the purpose for which we are created – TO WORSHIP HIM. But no, we chose the other route.

Romans 6:16: "Know ye not, that to whom ye yield yourselves servants to obey, his servants ye are to whom ye obey; whether of sin unto death, or of obedience unto righteousness?"

God is not watching to see if you will keep his rules. God knows you will keep his rules by just loving him. You aim to please whom you love. I had lived with the Lord for over 3½ years when I became frightened about a lot of demands people were putting on me. I was sick and could do nothing to pay the bills. I immediately ran to Jesus' lap. Fear is the opposite of faith and it is a sin. As I stained Jesus's robe, the Father asked who I was and Jesus told him it was Julie. I was convicted and as I stepped down, Jesus flicked off the stain. While sin is a stain, just like your baby makes a mess, we clean it up. Similarly, the blood of Jesus cleans up our sins. For those who have chosen a life outside Christ, they only know sin because no one can live righteous without God's help.

> 5*You meet him who rejoices in doing righteousness, who remembers you in your ways. Behold, you were angry, for we sinned- we continued in them a long time; and shall we be saved?* 6*For all of us have become like one who is unclean, and all our righteous deeds are like a filthy garment; and all of us wither like a leaf, and our iniquities, like the wind, take us away.* 7*There is no one who calls on your name, who arouses himself to take hold of you; for you have hidden your face from us and have delivered us into the power of our iniquities.* ***(Isaiah 64:6-7).***

Psalm 92:12: The righteous will flourish like a palm tree

John 1:1-18

1In the beginning was the Word, and the Word was with God, and the Word was God. 2The same was in the beginning with God. 3All things were made by him; and without him was not anything made that was made. 4In him was life; and the life was the light of men. 5And the light shineth in darkness; and the darkness comprehended it not. 6There was a man sent from God, whose name was John. 7The same came for a witness, to bear witness of the Light, that all men through him might believe. 8He was not that Light, but was sent to bear witness of that Light. 9That was the true Light, which lighteth every man that cometh into the world. 10He was in the world, and the world was made by him, and the world knew him not. 11He came unto his own, and his own received him not. 12But as many as received him, to them gave the power to become the sons of God, even to them that believe on his name. 13Which were born, not of blood, nor of the will of the flesh, nor of the will of man, but of God. 14And the Word was made flesh, and dwelt among us, (and we beheld his glory, the glory as of the only begotten of the Father,) full of grace and truth. 15John bare witness of him, and cried, saying, this was he of whom I spake, He that cometh after me is preferred before me: for he was before me. 16And of his fulness have all we received, and grace for grace. 17For the law was given by Moses, but grace and truth came by Jesus Christ. 18No man hath seen God at any time, the only begotten Son, **which is in the bosom of the Father**, he hath declared him.

"Which is in the bosom of the Father"

On this fateful day, I climbed on the Father like I was blindfolded, so as not to be consumed. The Lord has spoken to Ms. Laurence that I need to sit on His lap and talk. Sister Silvia also told me she saw me at the foot of the Father. I guess I mustered the strength to visit.

One day, as I opened my eye, and there was God over my house which has no need for walls. During this encounter, I opened my eyes and saw the throne and the host of God and heaven. I was in such awe to know what to say and reverential fear to share the experience. When they call God the rock of ages, He indeed is. The process is like climbing until I am asked to stop or welcomed. To my surprise, this did not happen when I got to the bosom. Jesus Christ came out of God's heart and told the Father to visit Julie (me). Suddenly, we went out to the bench by the stream where most people after a near-death experience claim to go first. We dialogued as I went into the water and back to my room to do my daily job as though nothing happened. Psalm 91 speaks of the shadow of the almighty. I have also encountered that, but this was the secret place of the Most High. I guess that is why I was blindfolded.

Faith is judging God faithful. Faith is expressing your trust that God will not lie and is reliable. Such trust will bring into your expectation that which you want. This is followed by the ministering angels bringing your desire to you. **Faith is a spirit** and **spirits are beings**. Therefore, **faith must be fed** so it can **grow as a person grows**.

Spirit of Faith.

In 2 Corinthians 4:13, it is written: "I believed; therefore, I have spoken." Since we have that same spirit of faith, we should also believe and speak. The Hebrew word *Ruach* means "wind." The being of the Holy Spirit or the person of the Holy Spirit is interchangeable because 'being' means 'to Exist'. This means the spirit of faith exists as a person exists. The Bible teaches that after the creation of man.

Who are the sons of God?

Children often ask, "What was there before God made the world?" The answer most adults would give is that God was there. That's true, but incomplete. God had company, and I'm not talking about the other members of the Trinity. The biblical answer is that the heavenly host was with God before creation. In fact, they witnessed it. What God says to Job in Job 38:4–7 is clear on that point:

"Where were you at my laying the foundation of the earth? Tell me, if you possess understanding. Who determined its measurement? Yes, you do know. Or who stretched the measuring line upon it? On what were its bases sunk? Or who laid its cornerstone, ***when the morning stars were singing together*** *and all the sons of God shouted for joy?"*

When God laid the foundations of the earth, the "sons of God" were there, shouting for joy. But who were these sons of God? Obviously, they weren't humans. This was before the creation of the world. We might think of them as angels, but that wouldn't be quite correct. These beings are lower than angels.

Chapter 7

THE HOLY HOST: GOD'S MORNING STARS

The unseen world has a hierarchy, something reflected in such terms as archangel versus angel. That hierarchy is sometimes difficult for us to discern in the Old Testament, since we aren't accustomed to viewing the unseen world like a dynastic household (more on that following), as an Israelite would have processed certain terms used to describe the hierarchy. In the ancient Semitic world, "sons of God" (Hebrew: *beney Elohim*) is a phrase used to identify divine beings with higher-level of responsibilities or jurisdictions. The term angel (Hebrew: *mal'ak*) describes an important but still lesser task – delivering messages.

In **Job 38:7**, the sons of God are referred to as "morning stars." The same description is found outside the Bible in ancient texts from the biblical world. Ancient people thought the stars were living entities. Their reasoning was simple: Many stars moved. That was a sign of life to the ancient mind. Stars were the shining glory of living beings from ancient days.

Matthew 2:1-2: Now after Jesus was born in Bethlehem of Judea in the days of Herod the king, behold, wise men from the east came to Jerusalem, saying, "Where is he who has been born king of the Jews? For we saw his star when it rose and have come to worship him."

Numbers 24:17: I see him, but not now; I behold him, but not near: a star shall come out of Jacob, and a scepter shall rise out of Israel; it shall crush the forehead of Moab and break down all the sons of Sheth.

Matthew 2:11: And going into the house they saw the child with Mary his mother, and they fell down and worshiped him. Then, opening their treasures, they offered him gifts, gold and frankincense and myrrh.

The stars also inhabited the divine realm—literally, in the sense that they existed off the earth. The ancients believed that divine beings lived far away from humans, in remote places where human habitation wasn't possible. The most remote place from this earth is the sky, the heavens.

Morning stars are the stars one sees over the horizon just before the sun appears in the morning. They signal a new life—a new day. The label works. It conveys the right thought. The original morning stars, the sons of God, saw the beginning of life as we know it—the creation of the earth. It is not surprising that in order for reproduction to give new life, a light has to shine. In the race for life –as with the spiritual race–the overcomer gets a light to start the next journey.

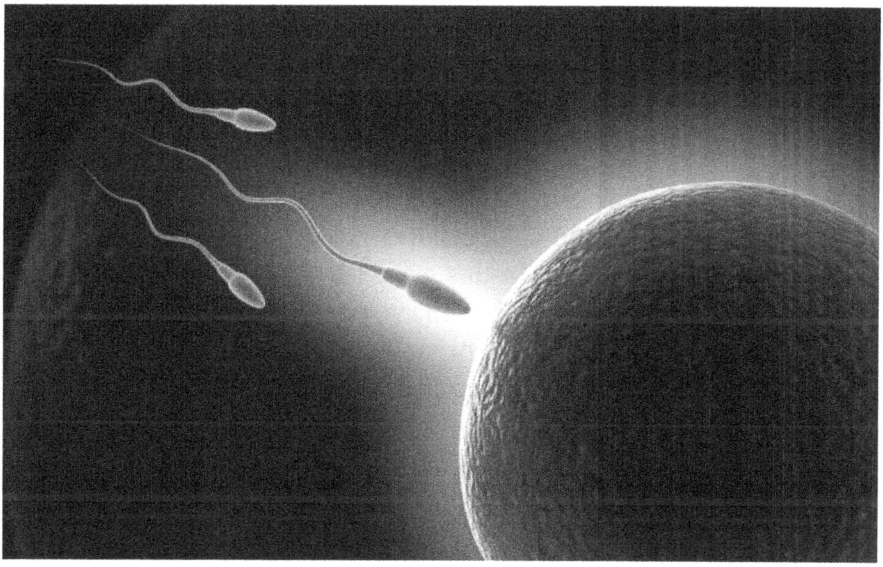

27TH APRIL, 2016

For the first time ever, scientists have captured images of the flash of light that sparks at the very moment a human sperm cell makes contact with an egg. The phenomenon has been observed in animals before, but no one has ever seen the spark of human conception. And what's even more incredible is the fact that some eggs burn brighter than others, which is a direct indication of their ability to develop into a healthy embryo, a team from Northwestern University found. So why do sparks literally fly at the moment of conception? Back in 2011, the Northwestern team discovered that sparks of zinc exploded at the point of conception in mice.

It took them a few years to figure out how to image this event, but by 2014, they'd managed to film the event for the first time ever and watched as billions of zinc atoms were released at the exact moment when a mammal's egg is pierced by a sperm cell. Using a new fluorescent sensor that is able to track the movements of zinc in live cells, the team caught a glimpse of an egg's zinc-storage capabilities, and found some 8,000 zinc compartments, each one containing around 1 million zinc atoms, just ripe for exploding. The tiny 'fireworks' that result was found to last for about 2 hours after fertilization.

Now, the same team has managed to film this event occurring in a human egg at the point of conception. "It was remarkable," says one of the team, Teresa Woodruff. "We discovered the zinc spark just five years ago in the mouse, and to see the zinc radiate out in a burst from each human egg was breathtaking."

If you follow health science news closely, you'll know that the vast majority of studies - even the most promising ones - fail to live up to their potential. In this case, the positive results in mice might not have been found in humans, or perhaps lab-based studies couldn't be replicated in living organisms. The researchers could've simply ran out of funding. But in this case, we can actually see how quickly the team has progressed from mice to humans, and in the space of just five years, figured out something that could change a lot of lives - the flashes of light that have now been seen in humans for the first time can be used to determine the viability of an individual egg. For couples relying on IVF treatments, that's huge, because around 50% of fertilized eggs do not develop properly due to some kind of unavoidable genetic mix-up.

Northwestern University

"This is an important discovery because it may give us a non-invasive and easily visible way to assess the health of an egg and eventually an embryo before implantation," says one of the researchers, Eve Feinberg. "There are no tools currently available that tell us if it's a good quality egg," she adds. "Often we don't know whether the egg or embryo is truly viable until we see if a pregnancy ensues. That's the reason this is so transformative. If we have the ability up front to see what is a good egg and what's not, it will help us know which embryo to transfer, avoid a lot of heartache, and achieve pregnancy much more quickly."

For the experiment, the team used a sperm enzyme to activate the egg (fertilizing the egg with actual sperm for research purposes is illegal under US federal law) and watched as the event triggered calcium levels to rise inside the egg, which prompted the release of zinc. "As the zinc shoots out, it binds to small molecules which emit a fluorescence which can be picked up by camera microscopes," Sarah Knapton explains for The Telegraph.

So why is zinc so special? The researchers found that eggs compartmentalize and distribute zinc to control the development of a healthy embryo - the mineral plays a vital role in controlling the 'decision' the egg makes to grow into an embryo - so the more zinc that's released, the brighter the flash, and the more viable the egg.

"The egg first has to stockpile zinc and then must release some of the zinc to successfully navigate maturation, fertilization, and the start of embryogenesis," said one of the team, Thomas O'Halloran, back in 2014. "But exactly how much zinc is involved in this remarkable process and where is it in the cell?" This is hopefully a question the researchers are now equipped to answer, and we can't wait to see what they discover next. The results have been published in Scientific Reports, and you can watch the footage below:

https://www.emory.edu/EMORY_REPORT/erarchive/1998/September/erseptember.21/9_21_98Size.htm

John Stone Street said it best. God's active act of creation, as described in Genesis 1, begins with those familiar words "Let there be light," or, as my Latin-loving colleagues prefer, "fiat lux."

Throughout the Scriptures, God's presence and power is associated with light. This is most obviously true in all of the writings of the Apostle John. In fact, as 1 John 1:5 tells us, "God is light and in Him there is no darkness at all." But this is more than history and metaphor. As it turns out, it's observable in the microscopic realm as well.

I have a severe eye problem lately, because of constantly being in God's presence. Like brother Paul, sometimes it seems God has allowed some physical challenge to remind me that as a child of God, I must remain humble.

Beautifully and Wonderfully Made

Psalm 139:14: I praise you because I am fearfully and wonderfully made; your works are wonderful, I know that full well.

God wasn't alone in the beginning.

God's Divine Family

God wants to grow His divine family with those that look and act like Him. This spirit loves to be with Him, and it never dies. God made us in His image and likeness and wants us to be filled with His Holy Spirit because there are a lot of them. Some examples are listed below:

What other name(s) is the Holy Spirit given in the Bible?

1. Spirit of glory (1 Peter 4:14)
2. Lord (1 Thessalonians 3:5)
3. God (Acts 5:3-4)
4. Spirit of revelation (Ephesians 1:17)
5. Spirit of the Son (Galatians 4:6)
6. Spirit of God (Genesis 1:2; 1 Corinthians 2:11; Job 33:4)
7. Eternal Spirit (Hebrews 9:14)
8. Spirit of the Lord (Isaiah 11:2; Acts 5:9)
9. Spirit of wisdom (Isaiah 11:2; Ephesians 1:17)
10. Spirit of counsel (Isaiah 11:2)
11. Spirit of might (Isaiah 11:2)
12. Spirit of understanding (Isaiah 11:2)
13. Spirit of knowledge (Isaiah 11:2)
14. Spirit of the fear of the Lord (Isaiah 11:2)
15. Spirit of judgment (Isaiah 4:4; 28:6)
16. Spirit of burning (Isaiah 4:4)
17. Spirit of the Lord God (Isaiah 61:1)
18. Breath of the Almighty (Job 33:4)
19. Comforter (John 14:16, 26; 15:26)
20. Spirit of truth (John 14:17; 15:26)
21. Power of the Highest (Luke 1:35)
22. Spirit of the Father (Matthew 10:20)

23. Spirit (Matthew 4:1; John 3:6; 1 Timothy 4:1)
24. Good Spirit (Nehemiah 9:20; Psalms 143:10)
25. Holy Spirit (Psalm 51:11; Luke 11:13; Ephesians 1:13; 4:30)
26. Free Spirit (Psalm 51:12)
27. Spirit of prophecy (Revelation 19:10)
28. Seven Spirits of God (Revelation 1:4)
29. Spirit of holiness (Romans 1:4)
30. Spirit of adoption (Romans 8:15)
31. Spirit of life (Romans 8:2; Revelation 11:11)
32. Spirit of Christ (Romans 8:9; 1 Peter 1:11)
33. Spirit of grace (Zechariah 12:10; Hebrews 10:29)

The Holy Spirit has characteristics called the fruit of the spirit.

Galatians 5:22-23: But the fruit of the Spirit is love, joy, peace, longsuffering, gentleness, goodness, faith, meekness, temperance: against such there is no law.

You may ask if all these spirits can exist in a person and the answer is yes. Just like your body parts, spirits are air and only act when asked to. Jesus said a spirit is known by their works. Good spirits do good works.

Therefore, from the beginning, God had company—other divine beings, the sons of God. Most discussions of what was around before creation omit the members of the heavenly host. That is unfortunate, because God and the sons of God, the divine family, are the first pieces of the mosaic. We've barely made it to creation so far, and already we've uncovered some important truths from the Bible that have the potential to affect our theology in simple but profound ways. Their importance, if it isn't clear yet, becomes quickly apparent.

As a Prophet and a Seer, I have witnessed these beings float when I pray to resurrect the dead, I have watched their network in getting some tasks done. Some of these beings have lower ranks and obey others and I see some manifest on an authoritative declaration under anointing and healing service communally or alone. I am not a wordsmith. I am a child of God by grace that has been instructed to testify of only what I have been shown and taught by God himself. I cannot accept that God was alone before He created the World .

I chose to borrow Michael S. Heiser's blog because his explanation on the host was what God allowed me to see. But these were not the sons of God, because sons of God had flesh and were marrying men.

Genesis 6:1-2: And it came to pass, when men began to multiply on the face of the earth, and daughters were born unto them, That the sons of God saw the daughters of men that they were fair; and they took them wives of all which they chose.

These beings had no body, wind with light combined. The wind in man was from fire. The Lord your God is spirit and Life.

Genesis 2:7: Then the LORD God formed a man from the dust of the ground and breathed into his nostrils the breath of life, and the man became a living being.

Job 33:4: The Spirit of God has made me; the breath of the Almighty gives me life.

Job 27:3: as long as I have life within me, the breath of God in my nostrils,

John 20:22: And with that he breathed on them and said, "Receive the Holy Spirit.

Job 34:14-15: If it were his intention and he withdrew his spirit and breath, all humanity would perish together and mankind would return to the dust.

Breath of Life

Physically, I have been visited by both God, the Father and Son, and will fervently proclaim that although God does the extraordinary by revealing Himself to me as He did with the prophets of old and some in the New Testament, He does not make himself physically accessible to everybody. He has instituted rules and laws to manage the billion planets and universes He created. While miracles like mine, Elijah and Daniel happened, it is still very rare. God is everywhere at every time in all the planets in the universe. I say this as a witness that he has given this uncommon privilege.

Ezekiel 37:9: Then he said to me, "Prophesy to the breath; prophesy, son of man, and say to it, 'This is what the Sovereign LORD says: Come, breath, from the four winds and breathe into these slain, that they may live."

According to this, God said our life in the breath is in the four corners of the earth and we can speak it into existence. The spirits of God were present on earth before creation, and they are the spirits I have listed earlier (I will elaborate more on their functions and workings in another book). When we put those spirits to sleep (because the spirits of God cannot work against God), we die in spirit.

Ephesians said he was one of the spirits before the creation of the universe.

Ephesians 1:3-6: [3]Praise be to the God and Father of our Lord Jesus Christ, who has blessed us in the heavenly realms with every spiritual blessing in Christ. [4] For he chose us in him before the creation of the world to be holy and blameless in his sight. In love [5] he predestined us for adoption to sonship through Jesus Christ, in accordance with his pleasure and will— [6] to the praise of his glorious grace, which he has freely given us in the One he loves.

Jesus needed the spirit to be awoken. He wanted more to be given and that is why he said we must be filled with the Holy Spirit. He called it forth before he left by prophecy:

John 20:22: And with that he breathed on them and said, "Receive the Holy Spirit.

Job 33:4: The Spirit of God has made me; the breath of the Almighty gives me life.

Job 27:3: as long as I have life within me, the breath of God in my nostrils,

Job 34:14-15: [14]If it were his intention and he withdrew his spirit and breath, [15]all humanity would perish together and mankind would return to the dust.

John 6:63 - It is the Spirit who gives life; the flesh profits nothing. The words that I speak to you are spirit, and they are life.

After God created man, he brought them into being when he breathed in them.

Matthew 5:14-16 (KJV): "[14]Ye are the light of the world. A city that is set on an hill cannot be hid. [15] Neither do men light a candle, and put it under a bushel, but on a candlestick; and it giveth light unto all that are in the house. [16] Let your light so shine before men, that they may see your good works, and glorify your Father which is in heaven."

The Lord is a Consuming Fire

Deuteronomy 9:3: But understand that today the LORD your God goes across ahead of you as a consuming fire; He will destroy them and subdue them before you. And you will drive them out and annihilate them swiftly, as the LORD has promised you.

Hebrews 12:28-29: [28]Therefore, since we are receiving an unshakable kingdom, let us be filled with gratitude, and so worship God acceptably with reverence and awe. [29]For our God is a consuming fire.

Psalm 104:4: He makes winds his messengers, flames of fire his servants.

Isaiah 63:10: Yet they rebelled and grieved his Holy Spirit. So he turned and became their enemy and he himself fought against them.

Matthew 12:31: And so I tell you, every kind of sin and slander can be forgiven, but blasphemy against the Spirit will not be forgiven.

Acts 7:51: "You stiff-necked people! Your hearts and ears are still uncircumcised. You are just like your ancestors: You always resist the Holy Spirit!"

Ephesians 4:30: And do not grieve the Holy Spirit of God, with whom you were sealed for the day of redemption.

Hebrews 10:29: How much more severely do you think someone deserves to be punished who has trampled the Son of God underfoot, who has treated as an unholy thing the blood of the covenant that sanctified them, and who has insulted the Spirit of grace?

The Holy Spirit has intentions, shows willfulness and discretion, loves, communicates, testifies, teaches, and prays (Neh. 9:20; Jhn 15:26; Acts 13:2; Rom. 8:26,27; 15:30; 1Cor. 12:11). These are qualities that distinguish Him as a person.

Nehemiah 9:20: You gave your good Spirit to instruct them. You did not withhold your manna from their mouths, and you gave them water for their thirst.

John 15:26: When the Advocate comes, whom I will send to you from the Father—the Spirit of truth who goes out from the Father—he will testify about me.

Acts 13:2: While they were worshiping the Lord and fasting, the Holy Spirit said, "Set apart for me Barnabas and Saul for the work to which I have called them."

Romans 8:26-27: In the same way, the Spirit helps us in our weakness. We do not know what we ought to pray for, but the Spirit himself intercedes for us through wordless groans. And he who searches our hearts knows the mind of the Spirit, because the Spirit intercedes for God's people in accordance with the will of God.

Romans 15:30: I urge you, brothers and sisters, by our Lord Jesus Christ and by the love of the Spirit, to join me in my struggle by praying to God for me.

1 Corinthians 12:11: All these are the work of one and the same Spirit, and he distributes them to each one, just as he determines.

Because faith grows. It must be fed. Because it is a being it must be exercised; thus faith works

John 14:1: Very truly I tell you, whoever believes in me will do the works I have been doing, and they will do even greater things than these, because I am going to the Father

John 6:29: Jesus answered, "The work of God is this: to believe in the one He has sent."

The root of the word faith is trust and the root of the word trust is truth. The reason we can trust God is because He is always true and faithful–He is full of truth and therefore reliable. I often play with God and He lets me. He draws a line with me always when He says something. I always make sure to stop what I'm doing and obey or repent.

Hebrews 11:1: Now faith is the assurance of things hoped for, the conviction of things not seen.

Romans 10:17 (NKJV): So then faith comes by hearing, and hearing by the word of God.

Matthew 21:22: And whatever you ask in prayer, you will receive, if you have faith.

Chapter 8

THE CHERRY AND ITS SEEDS

Revelation 2:7: He who has an ear, let him hear what the Spirit says to the churches. To the one who is victorious, I will grant the right to eat from the tree of life in the paradise of God.

On one of my early visits to heaven, I was given a cherry. I don't know why but I have always enjoyed cherry, particularly pickled cherry. I thought that was a welcome thing in heaven or a means to teach me how faith operated in heaven. I was not going to talk about it in this book; I shared it with my prayer partners and had little understanding of what they thought. I want you to understand that Jesus promised a new heaven free from sadness, stress, sickness and all the problems that we have here on earth.

Revelation 21:3-5: A New Heaven and a New Earth

[3]And I heard a loud voice from the throne saying: "Behold, the
dwelling place of God is with man, and He will live with them. They
will be His people, and God Himself will be with them as their God.
[4]He will wipe away every tear from their eyes, and there will be no
more death or mourning or crying or pain, for the former things have
passed away." [5]And the One seated on the throne said, "Behold, I
make all things new." Then He said, "Write this down, for these
words are faithful and true."

The Unscrambling of Lemons of Life into

Melons

Ephesians 1:4: For He chose us in Him before the foundation of the world to be holy and blameless in His presence.

We need a cure for the poisonous fruit we have eaten. Without wisdom first, knowledge is poison to the one who acquires it. Water is H_2O and is life to all. It contains two hydrogen molecules and one oxygen molecule. If we add one more molecule of oxygen, we will have H_2O_2 (hydrogen peroxide). Drinking higher concentrations of hydrogen peroxide can be very dangerous because it can cause tissue burns, which can lead to death. Similarly, not having the right balance between knowledge and wisdom eventually kills.

Reconcile with God – give and surrender your life to God through Jesus Christ.

What does it mean to accept Jesus as your personal Savior?"

Answer: To properly understand this question, you must first understand the terms "Jesus Christ," "personal," and "Savior."

Who is Jesus Christ?

Many people will acknowledge Jesus Christ as a good man, a great teacher, or even a prophet of God. These things are definitely true of Jesus, but they do not fully define who He truly is. The Bible tells us that Jesus is God in the flesh, God in human form (see John 1:1, 14). God came to the earth to teach us, heal us, correct us, forgive us, and die for us! Jesus Christ is God, the Creator, and the sovereign Lord. Have you accepted Jesus?

What is a Savior, and why do we need a Savior?

The Bible tells us that we have all sinned; we have all committed evil acts (Romans 3:10- 18). As a result of our sin, we deserve God's anger and judgment. The only just punishment for sins committed against an infinite and eternal God is an infinite punishment (Romans 6:23; Revelation 20:11-15). That is why we need a Savior!

Jesus Christ came to earth and died in our place. Jesus' death was an infinite payment for our sins (2 Corinthians 5:21). Jesus died to pay the penalty for our sins (Romans 5:8). Jesus paid the price so that we would not have to. Jesus' resurrection from the dead proved that His death was sufficient to pay the penalty for our sins. That is why Jesus is the one and only Savior (John 14:6; Acts 4:12)! Are you trusting in Jesus as your Savior?

Is Jesus your "personal" Savior?

Many people view Christianity as attending church, performing rituals, and/or not committing certain sins. That is not Christianity. True Christianity is a personal relationship with Jesus Christ. Accepting Jesus as your personal Savior means placing your own personal faith and trust in Him. No one is saved by the faith of others. No one is forgiven by doing certain deeds. The only way to be saved is to personally accept Jesus as your Savior, trusting in His death as the payment for your sins and His resurrection as your guarantee of eternal life (John 3:16). Is Jesus personally your Savior?

If you want to accept Jesus Christ as your personal Savior, say the following words to God. Remember, saying this prayer or any other prayer will not save you. Only believing in Jesus Christ and His finished work on the cross for you can save you from sin. This prayer is simply a way to express to God your faith in Him and thank Him for making provision for your salvation.

"God, I know that I have sinned against You and deserve punishment. but I believe Jesus Christ took the punishment I deserve so that through faith in Him I could be forgiven. I receive Your offer of forgiveness and place my trust in You for salvation. I accept Jesus as my personal Savior! Thank You for Your wonderful grace, Your forgiveness, and the gift of eternal life! Amen!"

Trusting in Elohim: He is Your Creator

Genesis 1:1: In the beginning God created the heavens and the earth. (NAS) (Gen 5:1)

The Bible begins with this simple but profound declaration

of faith – in the beginning, there was only God, nothing else. Everything else came to existence as the result of the activity of God. The first name for God that is found in the Bible in Genesis 1:1 is Elohim. What does this name mean? In the context of Genesis 1:1, it is a declaration that God (Elohim) alone is eternal. In a humanistic system, the universe becomes god-like, creating life out of lifeless matter through some mindless random system. By declaring that Elohim alone is eternal and is the Creator of all that exists, scripture gives dignity and meaning to our life and existence. We can know that life has meaning and purpose. If we are to discover that purpose, we must know the one who alone is eternal, our Creator. Further, it lays open the possibility that we can have eternal life. Mankind has always sought to find the secret to life after death, to eternal life, but it's no secret! Eternal life can only be found in a relationship to God, to Elohim, who is eternal.

The word "El" means mighty, strong, and prominent. The name "El" is most frequently found in compound names such as El Shaddai or El Elyon. In Numbers 23:22, God is spoken of as the "El" who brought Israel up out of Egypt. In Deuteronomy 10:17, we read that "Jehovah (Yahweh) your God (Elohim) is God of gods, and Lord of lords, the God (El) who is great, mighty, and dreadful." It was under the name "El" that God made great and mighty promises to Abraham and Jacob (Gen 17:1, 35:11).

Thus, we learn that when God uses the name Elohim, He is expressing His greatness and glory. The name Elohim contains the idea of creative and governing power, of omnipotence and sovereignty. We can clearly see this from Genesis 1:1 and 2:4. It is the Elohim who byHis mighty power created the vast universe; who says, and it is done; who brings into being what was not; by whose Word the world was framed.

The first thing God wanted us to know about Himself is that He is the Creator of the heaven and the earth. It is most appropriate that by this name, Elohim, God should reveal Himself – bringing cosmos out of chaos, light out of darkness, habitation out of desolation, and life in His image. This is a critical component of our faith. Once we lose confidence in Elohim as the Creator of all things, then it is just a short journey to doubt Him as our Savior. All of the other attributes of God are based on this affirmation that He existed prior to, is independent of, and sovereign over His creation – that the world reflects His glory. Another word from which some say Elohim is derived is "Allah". Allah is said to mean "to declare or to swear". Thus, it is said to imply a covenant relationship. In either case, whether Elohim is derived from "EL" or "Allah" the idea of the omnipotence of God is expressed. To make a covenant implies the power and right to do so.

Please look at Genesis 17 and perhaps we can see both "El" and "Allah" in this passage. In verse 1: "I am the Almighty God (El-Shaddai); walk before me and be blameless." In verse 7, "I will establish My covenant between Me and you and your descendants after you in their generation, for an everlasting covenant, to be God (Elohim) to you and your descendants after you." Elohim is mighty and powerful and wants to be in a covenant relationship with them. Other passages that speak of a covenant and Elohim are Gen 6:18; Gen 9:15-16; Gen 5:24; and 1 Kings 8:23.

It is also interesting to note that the name Elohim is a plural word and used as a noun; yet the verb or adjective that follows is often in the singular. "I am Elohim and there is no Elohim beside me." (Deut. 32:39; Isa. 45:5, 22). However, sometimes Elohim refers to Himself as "Us" as in Genesis 1:26 ("Let Us make man in Our image"), Genesis 3:22 which speaks of man becoming like one of "Us", or Genesis 11:7 when God says: "Let Us go down and confound their language."

Colossians 1:16 tells us that by Him or in Him were all things created. But, based on what we have already studied, we know creation was an act of Elohim. Therefore, Christ is in the Elohim or Godhead. In Ephesians 1:3-14, we find all three persons of the Godhead: "The grace of the Lord Jesus Christ and the Love of God, and the fellowship of the Holy Spirit, be with you all." Other scholars argue that the plural Elohim may have simply an intensifying sense, indicating God's majesty in the fullness of His power.

How Great Are Your Works?

Psalm 92:11-13: [11]My eyes see the downfall of my enemies; my ears hear the wailing of my wicked foes. [12]The righteous will flourish like a palm tree, and grow like a cedar in Lebanon.

[13]Planted in the house of the LORD, they will flourish in the courts of our God.

You must be born again

The teaching ministry of Jesus drew both friends and critics. Some openly loved him while others were too afraid to make it known in public. Ironically, the Pharisees who were the experts in searching the scriptures had a hard time believing the teaching of Jesus. But even among the skeptical Pharisees, there was one that was almost convinced that Jesus was the Messiah. And he wanted to get more details but he was too afraid to approach Jesus in full sight of his peers – so he went to Jesus at night.

John 3:1-4 (NIV): [1] Now there was a Pharisee, a man named Nicodemus who was a member of the Jewish ruling council. [2] He came to Jesus at night and said, "Rabbi, we know that you are a teacher who has come from God. For no one could perform the signs you are doing if God were not with him."[3] Jesus replied, "Very truly I tell you, no one can see the kingdom of God unless they are born again." [4] "How can someone be born when they are old?" Nicodemus asked. "Surely they cannot enter a second time into their mother's womb to be born!"

Nicodemus was flabbergasted by the answer of Jesus – you must be born again. What did Jesus mean? Did he mean that he was to go back to his mother's womb and be born for the second time? Or was Jesus teaching about reincarnation? He was totally confused. But Jesus cleared the air for him. He was not asking him to go back to his mother's womb; He was only asking him to accept God's plan for the salvation of man.

Two births?

Everyone was born into the earth by their mother. That is the first birth. It is the natural way that God has planned for man to populate the earth. We see this clearly in Genesis 1:26-28 where God created man, male and female, blessed them, and commanded them to multiply and fill the earth. God's perfect will for man was for him to live in Eden and enjoy what God had provided for them. Unfortunately, man fell prey to the deception of the devil and the moment he sinned, there was only one thing God could do: kick the man out of Eden. Assuredly, the Bible says the wages of sin is death

(Romans 6:23). This is exactly what happened when the man was kicked out of Eden – he experienced a spiritual death.

But that was not the end of the story. The day Adam and Eve tasted the forbidden fruit is the day the wheels starting moving towards the execution of God's plan to redeem mankind. God started by introducing his plan to individuals like Noah and Abraham, moved on to form the Nation of Israel, used Judges and prophets to prepare the way and orchestrated the coming of Jesus as God incarnate. Jesus came to die on behalf of mankind so that through His death, man can enjoy eternal life.

2 Corinthians 5:21: God made him who had no sin to be sin for us, so that in him we might become the righteousness of God.

This means that being born again is a divine act which results in the impartation of eternal life on those who believe (2 Corinthians 5:17; Titus 3:5; 1 Peter 1:3; 1 John 2:29; 3:9; 4:7; 5:1-4, 18). According to John 1:12, being born again means becoming the children of God by trusting in the name of Christ Jesus. So the first birth was the natural birth through your mother but the second birth is the spiritual birth which comes by faith in Jesus Christ as Lord

But why must a person be born again to inherit eternal life?

Well, Apostle Paul reminds us that all have sinned (Romans 3:23) and that the wages of sin is death. Sinners are therefore dead (spiritually). But then in Ephesians 2:1, he reminds us that we were made alive, "who were dead in trespasses and sins" receiving the forgiveness of sin by faith in Christ Jesus is, therefore, likened to a rebirth.

How does it work?

Ephesians 2:8–9: "For it is by grace you have been saved, through faith—and this not from yourselves, it is the gift of God—not by works, so that no one can boast."

Being born again is a gift that we receive by faith. In the Old Testament, an emphasis was placed on keeping the law to attain righteousness but Jesus made it easier for us – by paying the ultimate price on the cross, he bridged the gap between us and God

and now. Only one thing is required for us to be saved – to believe in Jesus Christ.

The scripture above explains it well. We are not saved by works but by grace through faith. Salvation is a free gift that is given freely by God through Jesus to anyone who believes. This is an important concept to grasp because the devil often deceives people into thinking they are not good enough to be saved. Or sometimes, if you succumb to his temptations, he convinces you that God would never want you back. But Jesus addressed this in the parable of the prodigal son. Even after he had wasted his life and his father's money, the father was glad to receive him back home.

That is how God responds to a remorseful heart.

But that does not mean that you can take the grace of God for granted. It is in your best interest to live a holy life once you have received the forgiveness of sins. Living in habitual sin damages your relationship with God severely. God is a righteous and holy God and even though he loves sinners, he hates sin with a passion. You will therefore not enjoy intimacy with God as long as you are living in sin.

Not yet born again?

If you are yet to be born again, you can do it by trusting the prompting of God through his Holy Spirit. Here is a simple prayer you can make to be born again today:

"Lord Jesus, you have said in your Word that if I confess with my mouth and believe with my heart that you are Lord, I shall be saved. This day, I believe that you died for my sins and rose again and I confess you as Lord and savior. I am therefore born again in Jesus name, Amen!"

You are now born again! Congratulations and welcome to the family!

What Next?

Now that you are born again, you must:

Share your faith in Christ with others

Sharing your faith with others is the best gift you can give to other people after you become a disciple of Christ. As scary as the thought of talking to others about your new-found faith is, just think where you would be today if someone hadn't extended the same courtesy to you. Sharing your faith isn't hard at all – you can begin by telling them your personal testimony including how meeting Jesus has helped you to become a better person. In the words of the scripture, they overcame by the word of their testimony (Revelation 12:11).

Follow the example of Christ through water baptism

It is not enough to just believe in Christ – you need to also follow the model of Christ by participating in water baptism. This is usually organized in most churches, so obviously, you may want to be part of a church first. Water baptism is an important symbolism in which your old self dies and is buried underwater and a new you rises up.

Mark 16:16: Whoever believes and is baptized will be saved, but whoever does not believe will be condemned.

Acts 2:38: And Peter said to them, "Repent and be baptized every one of you in the name of Jesus Christ for the forgiveness of your sins, and you will receive the gift of the Holy Spirit.

Romans 6:4: We were buried therefore with him by baptism into death, in order that, just as Christ was raised from the dead by the glory of the Father, we too might walk in newness of life.

Receive and be filled with the Holy Spirit

After receiving Jesus as Lord, you need to be filled with the Holy Spirit. Jesus taught that the Holy Spirit would lead his disciples in all truth and that he would empower them to lead a victorious life on earth. You cannot, therefore, lead a victorious Christian life without the power of the Holy Spirit.

John 14:26: But the Helper, the Holy Spirit, whom the Father will send in my name, he will teach you all things and bring to your remembrance all that I have said to you.

John 16:7-11: Nevertheless, I tell you the truth: it is to your advantage that I go away, for if I do not go away, the Helper will not come to you. But if I go, I will send him to you. And when he comes, he will convict the world concerning sin and righteousness and judgment: concerning sin, because they do not believe in me; concerning righteousness, because I go to the Father, and you will see me no longer; concerning judgment, because the ruler of this world is judged.

Turn away from all manner of sin

Being saved is not just a matter of going to church – you need to turn away from all sin. We come to God's kingdom through repentance (Mathew 4:17) and repentance literally means to make a turn-around.

1 Cor. 15:34: Come back to your senses as you ought, and stop sinning; for there are some who are ignorant of God – I say this to your shame.Be part of a Bible-based church

The importance of joining a Bible-based church cannot be overemphasized. A church community is not only important for fellowship and spiritual growth, but it can also be a source of social fulfillment.

Hebrews 10:25: Let us not give up the habit of meeting together, as some are doing. Instead, let us encourage one another all the more since you see that the Day of the Lord is coming nearer.

Study and base your life on the Word of God

More than ever before, every Christian must make up his mind to be a good student of the word of God. There are a lot of deception and heresies floating around, and the only way to see through them depends on your conversance with the Word of God. You must therefore take the Joshua charge very seriously.

Joshua 1:8: Here it is: This book of the law shall not depart out of thy mouth, but thou shalt meditate therein day and night, that thou mayest observe to do according to all that is written therein: for then thou shalt make thy way prosperous, and then thou shalt have good success.

Commune with God daily through prayer

Just like you cannot live without breathing, your spiritual life needs prayer. That is why Apostle Paul admonishes us to pray without ceasing. (1 Thess. 5:16-18). Through prayer, you get to present your requests to your heavenly Father and you also get to hear from Him. Prayer is also a weapon of spiritual warfare through which you resist the devil (James 4:7) and he will flee from you.

Trust in the promises of God

However, tough the going gets, you must resolve to alays trust in God's promises. This will help you to keep sane even if everything else fails. When Job lost all his property and all his family in one day, he still kept his trust in the Lord for he said:

Though he slays me, yet will I trust in him: but I will maintain mine own ways before him - **Job 13:15**.

Chapter 9

THE NEW JERUSALEM

How I wish I have the words to express what awaits us after the believer's judgment. When I requested for the New Jerusalem, a scroll was opened, and I had series of journeys. All I did was follow Him. I was like someone carried on the wings of the wind. However, due to the vastness of the universe, I couldn't really say where I was but all I know was that it's a wonderful place to behold.

Revelation 21: 9 – 27: [9]Then one of the seven angels with the seven bowls full of the seven final plagues came and said to me, "Come, I will show you the bride, the wife of the Lamb." [10]And he carried me away in the Spirit to a mountain great and high, and showed me the holy city of Jerusalem coming down out of heaven from God, [11]shining with the glory of God. Its radiance was like a most precious jewel, like a jasper, as clear as crystal. [12]The city had a great and high wall with twelve gates inscribed with the names of the twelve tribes of Israel, and twelve angels at the gates. [13]There were three gates on the east, three on the north, three on the south, and three on the west. [14]The wall of the city had twelve foundations bearing the names of the twelve apostles of the Lamb. [15]The angel who spoke with me had a golden measuring rod to measure the city and its gates and its walls. [16]The city lies foursquare, with its width the same as its length. And he measured the city with the rod, and all its dimensions were equal—12,000 stadiad in length and width and height. [17]And he measured its wall to be 144 cubits,e by the human measure the angel was using. [18]The wall was made of jasper, and the city itself of pure gold, as pure as glass. [19]The foundations of the city walls were adorned with every kind of precious stone: The first foundation was jasper, the second sapphire, the third chalcedony, the fourth emerald, [20]the fifth sardonyx, the sixth carnelian, the seventh chrysolite, the eighth beryl, the ninth topaz, the tenth chrysoprase, the eleventh jacinth, and the twelfth amethyst. 21And the twelve gates were twelve pearls, with each gate consisting of a single pearl. The main street of the city was pure gold, as pure as transparent glass. 22But I saw no temple in the city, because the Lord God Almighty and the Lamb are its temple. 23And the city has no need of sun or moon to shine on it, because the glory of God illuminates the city, and the Lamb is its lamp. 24By its light the nations will walk, and into it the kings of the earth will bring their glory. 25Its gates will never be shut at the end of the day, because there will be no night there. 26And into the city will be brought the glory and honor of the nations. 27But nothing unclean will ever enter it, nor anyone who practices an abomination or a lie, but only those whose names are written in the Lamb's Book of Life.

I rebuke Satan in the name of Jesus Christ to let go any child of God he is holding in any type of bondage. Loose their mind, joy, peace, finances, families, schools, home, businesses, vision, friendships, mission, health, marriage, and influences. The joy of the Lord is their strength; they will work in whole favor and peace of God now and always. In Jesus name.

The Vision of Joshua, the High Priest Zechariah 3:1-3

[1]Then the angel showed me Joshua the high priest standing before the angel of the LORD, with Satan standing at his right hand to accuse him. [2]And the LORD said to Satan: "The LORD rebukes you, Satan! Indeed, the LORD, who has chosen Jerusalem, rebukes you! Is not this man a firebrand snatched from the fire?" [3]Now Joshua was dressed in filthy garments as he stood before the angel.

Jude 1:9: But even the archangel Michael, when he was disputing with the devil about the body of Moses, did not himself dare to condemn him for slander but said, "The Lord rebuke you!"

Hebrews 3:6: But Christ is faithful as the Son over God's house. And we are His house, if we hold firmly to our confidence and the hope of which we boast.

Revelation 2:7: He who has an ear, let him hear what the Spirit says to the churches. To the one who is victorious, I will grant the right to eat from the tree of life in the paradise of God.

ROOT and STEM
of DAVID

Revelation 2:26–27: [26]And to the one who overcomes and
continues in my work until the end, I will give authority over
the nations. [27]He will rule them with an iron scepter and shatter
them like pottery— just as I have received authority from My
Father. [28]And I will give him the morning star.

Revelation 3:21: To the one who is victorious, I will grant the right to sit with Me on My throne, just as I overcame and sat down with My Father on His throne.

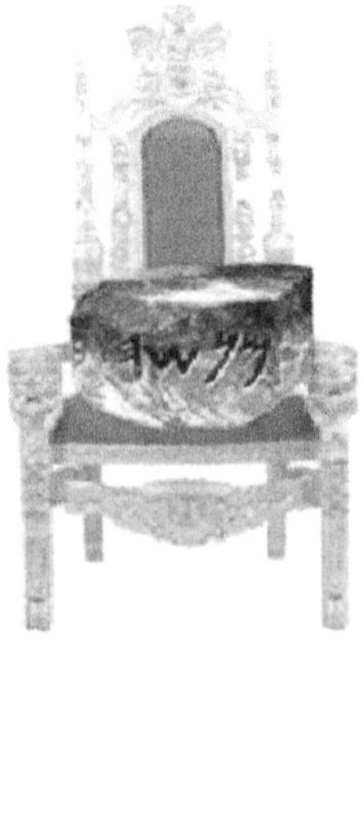

Daniel 7:18: But the saints of the Most High will receive the kingdom and possess it forever-- yes, forever and ever.

2 Timothy 4:8: Henceforth there is laid up for me a crown of righteousness, which the Lord, the righteous judge, shall give me at that day: and not to me only, but unto all them also that love his appearing.

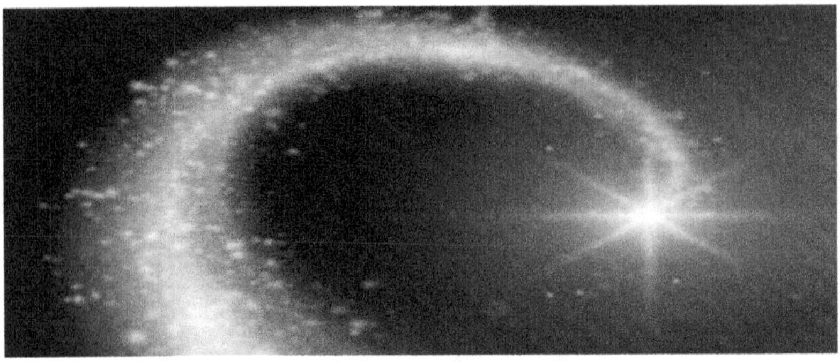

John 16:33: "I have told you these things, so **that** in me

you may have peace. In this world you will have trouble. But take heart! I have **overcome** the world."

Romans 12:21: "Do not be overcome by evil, but **overcome** evil with good."

1 John 4:4: "You, dear children, are from God and have overcome **them**, because the one who is in you is greater than the one who is in the world."

James 1:3-4: "because you know that the testing of your faith produces perseverance. Let perseverance finish its work so that you may be mature and complete, not lacking anything."

Revelation 2:7: "He who has an ear, let him hear what the Spirit says to the churches. To the one who is victorious, I will grant the right to eat from the tree of life in the paradise of God."

Chapter 10

JOY OF BECOMING A CHRISTIAN

What is the most important step you can take in life?

John 6:66–67: After this, a lot of his disciples left. They no longer wanted to be associated with him. Then Jesus gave the Twelve their chance: "Do you also want to leave?"

The most important step you can take in life is to follow Jesus. Jesus described himself as the way, the truth and the life (John 14:6). When you surrender to the Lordship of Christ, you begin to live with a purpose and more importantly, you get the assurance of salvation. In the verse above, some people were offended by him and left, but his disciples decided to make the most important decision: to keep following Christ. And the rest, as they say, is history.

According to Romans 3:23, what is the effect of sin?

The wages of sin is death. Death in this context means "separation from God." If you take a fish out of water, he dies. If you pluck a plant from the soil, it dies. Similarly, if you take a man from God's presence, he "dies." Even though you may still walk around doing your mundane tasks, you will be dead spiritually and will be what Apostle Paul referred to as a busy body (2 Thess. 2:11).

According to John 3:16, what happens when you chose Christ?

John 3:16 reveals the Love of God for mankind. It was this love that led him to send his only Son to come and die in our stead. By his death on the cross, we gain access to eternal life. So, when you accept Christ, the works of darkness are destroyed in your life and you begin a new walk with Him.

How do you get eternal life?

You do not get eternal life by working hard for it. You do not even get eternal life by being a good person – you only get eternal life by believing in Jesus as the son of God and accepting the work that he did on the cross.

Ephesians 2:8-9: [8]For by grace are ye saved through faith; and that not of yourselves: it is the gift of God: [9]Not of works, lest any man should boast.

What does it mean to be born again?

To be born again means accepting the Lord Jesus as your Lord and Savior and starting a new life. When that happens, you get ushered into the kingdom of God and become a son of God.

2 Cor. 5:17: Therefore, if any man be in Christ, he is a new creature: old things are passed away; behold, all things are become new.

John 1:12: But as many as received him, to them gave he power to become the sons of God, even to them that believe on his name.

Have you made the personal decision to forsake your sins and make Jesus your Lord and Savior?

What is the significance and importance of water baptism?

Water baptism signifies that we are dead to the old self and a new life has begun in Christ. The dipping in water signifies the burying of the dead (old) man and the emerging out of water typifies the resurrection with Christ from the dead.

Colossians 2:12: Having been buried with him in baptism, in which you were also raised with him through your faith in the working of God, who raised him from the dead.

Romans 6:4: We were therefore buried with him through

baptism into death in order that, just as Christ was raised from the dead through the glory of the Father, we too may live a new life.

Scripture says we are saved by grace. What does this mean?

Ephesians 2:8: [8]For by grace are ye saved through faith; and that not of yourselves: it is the gift of God.

Being saved by grace means that salvation is not your right or something you deserve. On the contrary, salvation is a free gift from God to mankind. It is available for anyone who will take it and there are no favorites in the eyes of God – anyone who is willing and obedient will be saved.

Why do you think a public confession of Christ is important?

Jesus taught that if we confess him in public, he will also confess us before God in heaven. Confessing Christ in public also helps establish accountability because you will always be conscious that people around you know you are now born again.

Matthew 10:32: Whoever acknowledges me before others, I will also acknowledge before my Father in heaven.

Transform your life to be like Christ- let the spirit lead

Question: "What is the meaning of 'from glory to glory' in 2 Corinthians 3:18?"

Answer: "But we all, with unveiled face, beholding as in a mirror the glory of the Lord, are being transformed into the same image from glory to glory, just as from the Lord, the Spirit" - 2 Corinthians 3:18 (NASB).

With those few words—"from glory to glory"—Paul sums up our entire Christian life, from redemption and sanctification on earth, to our glorious eternal welcome into heaven. There is a great deal of content packed into those few words. It's all so important that Paul labors at great length, from 2 Corinthians 2:14 through the end of chapter 5, to open his readers' eyes to a great truth. Let's see why that truth matters so much.

The same Greek word for "glory" is used twice in the phrase "from glory to glory"; yet each usage refers to something different. The first "glory" is that of the Old Covenant—the Law of Moses—while the second is that of the New Covenant, the gospel of Jesus Christ. Both have astonishing splendor. The Old Covenant was given to Moses directly from God, written by God's own finger (Exodus 31:18). That root of our Christian faith is glorious indeed; it's the glory we're coming "from." Yet the New Covenant, the glory we're going "to," far surpasses that of the Old.

The transformation is from the glory of the Law. Like the stone it was written on, the Law was inflexible and absolute, applying to all Israelites without much regard for individual circumstances (Hebrews 10:28). Though holy, good, and righteous in itself (Romans 7:12), the Law was, for us sinners, the letter that kills us (2 Corinthians 3:6). The Law was an external force to control behavior. In addition, stone, despite its strength, is earthly and will eventually wear away. The Law was merely a temporary guardian (Galatians 3:23–25) until something better came along.

The transformation is to the glory of the New Covenant, which far surpasses the Old in every way. It forgives us of our sin and gives us (sinners) life (John 6:63). It is written on believers' hearts by the Holy Spirit (Jeremiah 31:33; 2 Corinthians 3:3), so our obedience to God springs up from within us by God-given desires rather than by threats of legal punishment. In place of a cold set of writings as a guide for pleasing God, we now have Father, Son and Holy Spirit making their home with us, fellowshipping in loving intimacy, teaching us everything we must know and do (John 14:23; 16:13). That position in Christ is as permanent, eternal, and spiritual as God Himself, rather than temporary and earthly.

Paul is intent on directing Christians to focus on the spiritual glory of the New Covenant rather than the physical glory of the Old, as many Jews in his day refused to do. He compared the two types of glory by recalling how Moses absorbed and reflected God's glory for a time after being in his presence (2 Corinthians 3:7–11, 13; cf. Exodus 34:29–35). Though Moses' glow had a spiritual cause, there was nothing spiritual about the effect—any person, regardless of his relationship with God, could see the glow on Moses' face, which he covered with a veil.

This is not so with the glory of the New Covenant which can be seen only with a believer's spiritual eyes, which Paul is doing his best to open, so we can discern the gospel's glory. So, he writes, "For God, who said, 'Let light shine out of darkness,' made his light shine in our hearts to give us the light of the knowledge of God's glory displayed in the face of Christ" - 2 Corinthians 4:6.

But, as we move from glory to glory, there's something even more important about the glory of the New Covenant that Christians must understand: its supernatural power to transform us. This brings us to God's ultimate purpose and destination for every believer: to transform us into the image of his own beloved Son (2 Corinthians 3:18; Romans 8:28–30; Philippians 3:20–210.

Before he finishes with the topic of being transformed from glory to glory, Paul presents yet one more astonishing claim: "Therefore, if anyone is in Christ, he is a new creation; the old has gone, the new has come!" - 2 Corinthians 5:17.

This is the invitation the Lord makes to all Christians, to have our lives radically transformed here and now, by opening our eyes to see the glorious journey He is taking us on "from glory to glory."

Intimacy and worship of God

Question: "How can I experience true intimacy with God?"

Answer: True intimacy with God is something that has been sought by Christians since Christ walked the earth. It is natural for Christians to long to experience the closeness of an intimate relationship with God. However, true intimacy with God is not simply a feeling similar to a romantic relationship. It goes much deeper than emotion, down to our very souls and reflected by our actions. "For the LORD detests the perverse but takes the upright into his confidence" (Proverbs 3:32). God cannot have intimacy with evil or disobedient Christians. True intimacy with God begins with drawing near to Him.

God will never draw near to those who do not draw near to Him, and the way we "draw near" is through righteousness. "Draw near to God, and he will draw near to you. Cleanse your hands, you sinners, and purify your hearts, you double-minded" (James 4:8).

Certainly, God will never draw near in intimacy with the unrighteous, but with those who have been cleansed by the blood of Christ and have received His righteousness at the cross (2 Corinthians 5:21). They have the hope of intimacy with God. In fact, it is only those who have been saved by grace through faith (Ephesians 2:8–9) who have that hope, because Christ is the hope through which we draw near to God (Hebrews 7:19).

Jesus is, in fact, the model of intimacy with God because He and the Father are one (John 10:30), and no relationship can be closer than that oneness with the Father that Jesus experienced. His relationship with the Father was characterized by love and obedience. In love, Jesus came to earth to do His Father's will. He did nothing on His own, but in all things did the will of His Father (John 5:30). This was most evident in the Garden of Gethsemane the night before His crucifixion. Suffering the agony of anticipating what was to come, Jesus asked that the fate He was about to suffer might be removed from Him, but He ended the plea by saying, "Yet not My will, but Yours be done" (Luke 22:42). Here we see a perfect example of true intimacy reflected in obedience as Jesus yielded His will to that of His Father.

If we hope to attain true intimacy with God, Jesus must be our model. We love God because He first loved us (1 John 4:10), and we prove our love for Him by obeying Him. Jesus told His followers, "If you love me, keep my commandments" (John 14:15). When we obey Him and keep what He has commanded, He promised that we will remain in His love, just as He remains in the love of the Father by doing the Father's will (John 15:10). There can only be intimacy with God when we are in good fellowship with Him through obedience. Then we can know the joy and peace that comes from trusting Him and yielding to His will, just as Jesus did.

Rejoicing in Trials

James 1:6- 7 – 6But he must ask in faith, without doubting, because he who doubts is like a wave of the sea, blown and tossed by the wind. 7That man should not expect to receive anything from the Lord. 8He is a double-minded man, unstable in all his ways.

Romans 7:21 – So this is the principle I have discovered:

When I want to do good, evil is right there with me.

Romans 7:23-25: But I see another law at work in my body, warring against the law of my mind and holding me captive to the law of sin that dwells within me. What a wretched man I am! Who will rescue me from this body of death? Thanks be to God, through Jesus Christ our Lord! So then, with my mind I serve the law of God, but with my flesh I serve the law of sin.

Since life is short and eternity is forever, live by God's time, not by man's time.

Living in the Spirit

Romans 8:1-3: [1]Therefore there is now no condemnation for those who are in Christ Jesus. [2]For in Christ Jesus the law of the Spirit of life has set you free from the law of sin and death. [3]For what the Law was powerless to do in that it was weakened by the flesh, God did by sending His Own Son in the likeness of sinful man, as an offering for sin. He thus condemned sin in the flesh.

Ephesians 2:16: and reconciling both of them to God in one body through the cross, by which He extinguished their hostility.

Philippians 2:6-8: [6]Who, existing in the form of God, did not consider equality with God something to cling to, [7]but emptied Himself, taking the form of a servant, being made in human likeness. [8]And being found in appearance as a man, He humbled Himself and became obedient to death—even death on a cross.

Isaiah 42:1: "Behold, My Servant, whom I uphold; my chosen one in whom My soul delights. I have put My Spirit upon Him; He will bring forth justice to the nations."

Matthew 20:28: "just as the Son of Man did not come to be served, but to serve, and to give His life as a ransom for many."

John 1:14: The Word became flesh and made His dwelling among us. We have seen His glory, the glory of the one and only Son from the Father, full of grace and truth.

Romans 8:3: For what the Law was powerless to do in that it was weakened by the flesh, God did by sending His own Son in the likeness of sinful man, as an offering for sin. He thus condemned sin in the flesh,

2 Corinthians 8:9: For you know the grace of our Lord Jesus Christ, that though He was rich, yet for your sakes He became poor, so that you through His poverty might become rich.

2 Corinthians 13:4: For He was indeed crucified in weakness, yet He lives by God's power. And though we are weak in Him, yet by God's power we will live with Him to serve you.

Galatians 4:4: But when the time had fully come, God sent His Son, born of a woman, born under the Law,

Hebrews 2:17: So He had to be made like His brothers in every way, that He might become a merciful and faithful high priest in service to God, in order to make atonement for the sins of the people. But made himself of no reputation, and took on him the form of a servant, and was made in the likeness of men:

Revelation 22: 8–9: [8] I, John, am the one who heard and saw these things. And when I heard and saw, I fell down to worship at the feet of the angel who showed me these things. [9] But he said to me, "Do not do that. I am a fellow servant of yours and of your brethren the prophets and of those who heed the words of this book. Worship God!"

Heaven is your Home, Eternity is your Promise

First, I will introduce you to the Father who, by you believing in Him, will do two things: Transform your mind and renew your thinking. This may include painful separation of things you like that do not like you. This means things that are convenient for you to do but whose end will destroy you.

Put on Christ

Romans 13:14 – Rather, clothe yourselves with the Lord Jesus Christ, and do not think about how to gratify the desires of the flesh.

Do everything without grumbling

Philippians 2:12 – Therefore, my dear friends, as you have always obeyed—not only in my presence, but now much more in my absence—continue to work out your salvation with fear and trembling,

Romans 10:10 – For it is with your heart that you believe and are justified, and it is with your mouth that you profess your faith and are saved.

John 5:39 – Search the scriptures; for in them ye think ye have eternal life: and they are they which testify of me.

1 John 3:8 – He that committed sin is of the devil; for the devil sinneth from the beginning. For this purpose, the Son of God was manifested, that he might destroy the works of the devil.

God wants to help you by healing your sickness, be it physical or spiritual. All healing comes through Jesus Christ.

Encounter with Hell and Mega-Earth Kepler-10c

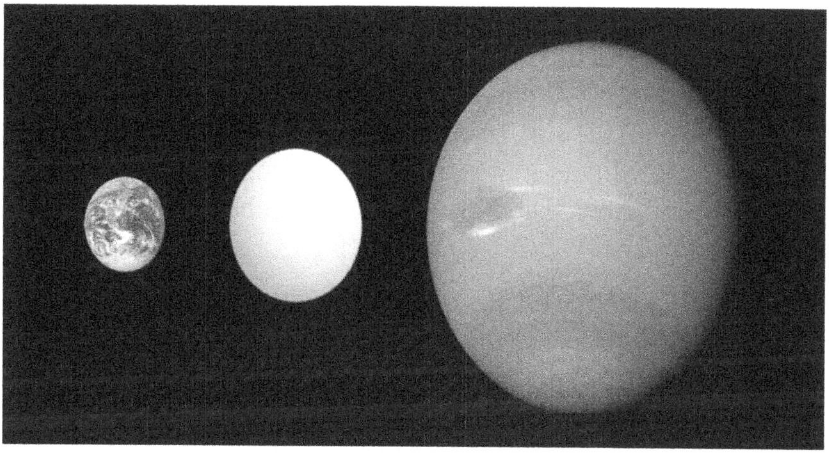

Astronomy finds earth ten times the size of earth.

Revelation 21:1 – Then I saw "a new heaven and a new earth," for the first heaven and the first earth had passed away, and there was no longer any sea.

Matthew 24:35: Heaven and earth will pass away, but my words will never pass away.

I do not know which one is worse: two thousand years before or after Jesus Christ has left the earth. Neither Christians nor scientists have considered that the Universe is God's playground. The Bible tells the Christians that those that follow Jesus to the end (overcoming this present world and its pressures) will reign with him. As God said in Revelation 2:28, "I will give him nations and the morning star."

Reign Who?

If human beings either go to Heaven or Hell, that will mean everyone in heaven would have overcome this world. Revelation 12:11 says "They triumphed over him by the blood of the Lamb and by the word of their testimony; they did not love their lives so much as to shrink from death." To "overcome" the trials and temptations of life and thus remain loyal to him to the end (see: Rev.2:7, 11, 17, 26; 3:5, 12, 21).

Revelation 22:16: I, Jesus, have sent mine angel to testify unto you these things in the churches. I am the root and the offspring of David, and the bright and morning star.

To the Church in Thyatira

Revelation 2:25-28: [25]Nevertheless, hold fast to what you have until I come. [26]And to the one who is victorious and continues in My work until the end, I will give authority over the nations. [27]He will rule them with an iron scepter and shatter them like pottery—just as I have received authority from My Father [28] I will give him the morning star.

Some scholars consider the reference of us getting the morning star to be Christ himself. The Bible says Jesus Christ is a morning star. The Lord will also give nations to the saints who overcome at the end. As I have shown by scientific discoveries, these stars can have universe or universes attached to them. Is it not mind-blowing to understand that just like the eyes cannot see the stars, planets, or their associated universes, the Father is preparing what eyes have not seen or ears have not heard for those who love him?

The biblical answer is that the heavenly host was with God before creation. In fact, they witnessed it. What God says to Job in Job 38:4–7 is clear on that point:

"Where were you at my laying the foundation of the earth? Tell me, if you possess understanding. Who determined its measurement? Yes, you do know. Or who stretched the measuring line upon it? On what were its bases sunk? Or who laid its cornerstone, when the morning stars were singing together and all the sons of God shouted for joy?"

In the book of Job, the morning star means the worshipers. The one that sings when a man becomes a saint.

The Order of Resurrection

1 Corinthians 15:21-23: [21]For since death came through a man, the resurrection of the dead comes also through a man. [22]For as in Adam all die, so in Christ all will be made alive. [23] But each in his own turn: Christ the first fruits; then at His coming, those who belong to Him…

"Jesus Christ is the 'morning star'" (2 Peter 1:19). Some

suggested this to be a mysterious promise God revealed from the beginning even to David in the book of psalms.

Psalm 2:8: Ask Me, and I will make the nations Your inheritance, the ends of the earth Your possession.

The Lord does not want us to remain babies; we must grow in trusting Him and gaining the ability to manage nations here on earth because when we go to heaven, we will get nations to manage.

Psalm 2:9: "You will break them an iron sceptre; You will shatter them like pottery."

He goes on to point out that this star symbolism reflects the Old Testament and inter-testament emphasis on the "celestial" or heavenly nature of the coming Messiah (cf. Num. 24:17; Mal. 4:2). We are in the kingdom now (Col. 1:13), but there is a more glorious phase we are yet to enter (2 Pet. 1:11). We enjoy salvation presently (Mk. 16:16), but there is a greater dimension – a heavenly salvation – to be received at death (2 Tim. 4:18). There is a sense in which we are with Christ now (Mt 18:20; 26:29; 28:20), but there is a more exalted state in which we will be with him ultimately (Phil. 1:23; 2 Cor. 5:8).

1 Corinthians 15:41: There is one glory of the sun, and another glory of the moon, and another glory of the stars: for one star different from another star in glory.

Revelation 2:28: And I will give him the morning star.

Morning Star

I suggest that you look up "Morning Star" in Wiktionary, the free dictionary. The phrase "Morning Star" may most likely refer to **astronomy:**

Morning star, most commonly used as a name for the planet Venus when it appears in the east before sunrise.

Morning star, a name for the star Sirius, which appears in the sky just before sunrise during the Dog Days.

Morning star, a (less common) name for the planet Mercury when it appears in the east before sunrise.

Matthew 2:10: When they saw the star, they rejoiced with exceeding great joy.

Matthew 2:2: Saying, where is he that is born King of the Jews? For we have seen his star in the east, and are come to worship him.

Your god, which ye made to yourselves {idols}

I remember God appearing to me to tell me how disappointed he gets when we create images or idols and compare them to Him. Exalting other things above our Creator is a terrible sin. I fell on my knees and prayed a prayer that could only come from the Holy Spirit within: “Father forgive them for they do not know what they are doing” This was one of my most frightening encounters. You never want to see God angry. I say this is from the Holy Ghost. Jesus also prayed this prayer at Calvary and Brother Steven prayed it at his stoning. I know that I have the righteousness of God but only by grace. I never knew that prayer was in me.

A Chosen People

Revelation 9:1: And the fifth angel sounded, and I saw a star fall from heaven unto the earth: and to him was given the key of the bottomless pit.

Genesis 15:5: He took him outside and said, "Look up at the sky and count the stars--if indeed you can count them." Then he said to him, "So shall your offspring be."

Revelation 1:6: and has made us to be a kingdom and priests to serve his God and Father—to him be glory and power for ever and ever! Amen.

Isaiah 61:6: But you shall be called the priests of the LORD; they shall speak of you as the ministers of our God; you shall eat the wealth of the nations, and in their glory you shall boast.

Exodus 19:6: And you shall be to me a kingdom of priests and a holy nation.’ These are the words that you shall speak to the people of Israel.”

Peter 2:9: But you are a chosen race, a royal priesthood, a holy nation, a people for his own possession, that you may proclaim the excellencies of him who called you out of darkness into his marvelous light.

The Need for Wisdom to Successfully Live

The excellence of wisdom was seen from the beginning before creation. It is said that the fear of God is the beginning of wisdom. The LORD created me as His first course, before His works of old.

Proverbs 8:23: [23]From everlasting I was established, from the beginning, before the earth began. [24]When there were no watery depths, I was brought forth, when no springs were overflowing with water.

John 5:26: For just as the Father has life in himself, thus he has granted the Son to have life in himself.

Mark 16:16: **"**He that believeth and is baptized shall be saved; but he that believeth not shall be damned."

Romans 14:23: "And he that doubteth is damned if he eat, because he eateth not of faith: for whatsoever is not of faith is sin".

Thessalonians 2:12: "That they all might be damned who believed not the truth, but had pleasure in unrighteousness".

The damnation of man speaks to the everlasting torment in hell, in the life after death as a sinner. No one is damned by God. Instead, only a human can condemn his/her soul to damnation. God's actions make it clear that He wants all humankind to be with Him forever; He wants man to be reconciled to Him. The reconciliation of man unto God is the most important reason and cause for salvation. The gift of salvation, whether accepted or refused by man, is the Birth, Death, Resurrection, Ascension and promised Return of Christ.

Therefore, when one decides to not believe in the opportunity to have everlasting life, then one automatically chooses everlasting death, which is damnation to hell. God devised the plan of salvation and made it available to all of humanity. In His plan, He also gave man the leverage to choose, rather than dictating. Love does not force itself on other people as evil spirits tend to do, and since God is love, He gives us free will.

While damnation is solely based on the decision of the human, it is also an automatic result when one decides not to believe in Christ and accept the gift of Salvation. The thief on the cross immediately condemned himself to damnation, simply because he refused forgiveness and an opportunity for everlasting life. He did not believe that Jesus was indeed the Christ. This scene at the cross could have happened and not been recorded, but this story made its way into the Bible so that we could understand the difference between believing and not believing and the results derived from either (eternal life and eternal damnation respectively). Again, here, God gives us the information we needed, to ensure we can make the best decision. Note, damnation for the unbeliever is not hell but the rejection of God.

Matthew 10:33: But whoever disowns me before others, I will disown before my Father in heaven.

Damnation means (in Christian belief) condemnation to eternal punishment in hell. Just like in a human court, no one can throw you in citation without trial. When your name is not found in the Lamb's Book of Life, it means that you have rejected Jesus and God. Before you are thrown in hell, you are shackled with everything that belongs to the devil, whom you chose as master by rejecting the gift of salvation. In my encounter with damnation, the damned person going to hell received every sickness, horror and torment of life before being directed to hell. God separated His life, which had been borrowed to the sinners, and every happy thing life gives to both the righteous and the unrighteous from those who are damned.

This damnation is part of the hell process, so it doesn't stop even in hell. In Jesus' words, on Judgment Day, we see several important truths: it's he who follows Jesus that gets saved (Matthew 7:21). Nominal Christianity cannot save. Also, it's not a demonstration of spiritual insight or power that saves (verse 22). A person can seem like a Christian in the eyes of other people, yet still be an "evildoer" in God's sight and sent away from His presence (verse 23). Only those who do the Father's will and who are known of God will enter heaven.

When Jesus said, "I never knew you," to the feigned disciples, He meant that He never recognized them as His true disciples or His friends. He never had anything in common with them nor approved of them. They were no relations of His (Mark 3:34–35). Christ did not dwell in their hearts (Ephesians 3:17), nor did they have His mind (1 Corinthians 2:16). In all these ways and more, Jesus never knew them. Note that Jesus is not breaking off the relationship here—there was never a relationship to break off. Despite their high-sounding words and showy displays of religious fervor, they had no intimacy with Christ.

So, it turns out that what matters isn't so much that we know God on some level, but that God knows us. As Paul explained, "Whoever loves God is known by God" (1 Corinthians 8:3; also see Galatians 4:9). The Lord "tends His flock like a shepherd" (Isaiah 40:11), and He knows who His sheep are (John 10:14).

The benefit of being alive in this life is that we have the opportunity to determine our future – everlasting life or everlasting damnation.

What is Everlasting Life?

Seven Places Jesus Bled to Set You Free

Hebrews 8:1-2: Now this is the main point of the things we are saying: We have such a High Priest, who is seated at the right hand of the throne of the Majesty in the heavens, a Minister of the sanctuary and of the true tabernacle which the Lord erected and not man.

The Lord Jesus Christ is our High Priest who has gone before us to make atonement for the sins of mankind and to reconcile us back to God. Our sins, iniquities, transgressions, generational curses, and sicknesses have been washed away by the cleansing power of the Savior's blood. Jesus bled seven times to set us free from the curse of Adam, giving us a complete salvation.

Adam sinned in the garden of Eden, the steps for our redemption started at another garden: the Garden of Gethsemane. At the Garden of Eden, man received judgement and Christ came as a man in the garden to take that curse and damnation of God from us. Man did not repent but continued in sin. This fallen relationship with God received the generational consequences; God extended it to the work of man's hand and other troubles in daily living. Sin, sickness, and death began to move down the bloodline of all humanity bringing destruction and devastation to everyone, sending all to hell that did not receive the cleansing blood of Christ.

The blood of Jesus Christ is the only cleansing agent that can wash away the sins of man. No other remedy can repair man's fall from grace, only the Savior's blood. Until man cries out for mercy, confesses his sinful condition, and repents of his iniquities, he will remain in an eternal position of being an enemy of God, destined for eternal damnation.

Romans 5:18: Therefore, as through one man's offense, judgment came to all men, resulting in condemnation; even so through one Man's righteous act the free gift came to all men, resulting in justification of life.

Chapter 11

REJECTION OF GOD PAID IN FULL

And being in agony, He prayed more earnestly. Then His sweat became like great drops of blood falling down to the ground. -**Luke 22:44**

Adam's sin expelled him from the garden of Eden and separated all of mankind from the grace of God. Jesus, the last Adam, gained entry for us all to once again live in God's presence by the blood that He shed in the Garden of Gethsemane.

Taking upon Himself the sins of the world by becoming the supreme sacrifice, in agony, He sweat blood to give us access once again to the Father's presence. He literally became sin in our place and by His blood delivered all of mankind from the penalty of death and eternal damnation.

2 Corinthians 5:21: For He made Him who knew no sin to be sin for us, that we might become the righteousness of God in Him.

Imagine this: The Passover supper has been eaten. Jesus has concluded His "upper room discourse," as recorded in John's gospel, including the high priestly prayer of Jesus for His disciples in chapter 17. Jesus and the disciples have sung a hymn, they have left the upper room, and they have crossed the Kidron to the Mount of Olives, specifically to the Garden of Gethsemane. Luke mentions only that the party went to the Mount of Olives, for his Gentile readers would not have known the precise location that some of the Jewish readers (of other gospels) would have recognized.

There is no wonder why our Lord was "sorrowful and troubled" (Matthew 26:37), and His soul was "overwhelmed with sorrow to the point of death" (Matthew 26:38). Jesus' agony was due to the cross which loomed before Him. He was not in agony because He would be forsaken by men, but that He would be forsaken and smitten by God. Jesus was dreading, suffering in the anticipation of His bearing of the sins of the world and the wrath of God which man deserved.

This text tells us that because Jesus bore the wrath of God (the "cup," as it were) in the sinner's place, it is not necessary for men to drink this cup as well. Salvation comes when a person comes to faith in Christ as the One who was innocent, and yet died in their place, bearing the wrath of God which their sins deserved. Those who reject Christ and His atoning sacrifice must bear the wrath of God, which will be poured out on unbelievers in the future. It is this wrath to which the Book of Revelation refers.

There are many disagreements among evangelicals as to when and how the Lord's return will come, but one thing seems certain to me, based on our text: No Christian will go through the Tribulation, the future outpouring of God's wrath upon an unbelieving world. All who are godly will suffer "tribulation" (small "t"), which is the wrath of unbelieving men toward God (cf. 2 Timothy 3:12), but the Great Tribulation (big "T")—the outpouring of divine wrath on sinful men—will only come upon the unbelieving. The Great Tribulation is a horrifying repeat of the agony of Calvary, which men must endure because of their rejection of the Savior, and it will only come upon unbelievers.

A Problem Passage

Luke 22:43–44: 43 An angel from heaven appeared to him and strengthened him. 44 And being in anguish, he prayed more earnestly, and his sweat was like drops of blood falling to the ground.

Luke 22:43-44 pose a problem for some. First, these verses are not found in a very few of the "older" manuscripts. Since "older" is not necessarily "better," and since only a few manuscripts omit these verses, I find it easy to assume that the verses are original. The very fact that these verses are difficult to understand and that they are not found in the parallel accounts is strong evidence for their originality, in my opinion.

TETELESTAI "IT IS FINISHED" "PAID IN FULL"

MEDITATION ON "IT IS FINISHED" (John 19:30-note)

Which can also be translated "PAID IN FULL!" THREE words in English, ONE word in Greek – TETELESTAI! The GREATEST WORD from the GREATEST MAN on the GREATEST DAY in all eternity! One word, but no word ever uttered has so changed the history and destiny of mankind. In Latin tetelestai is rendered with two words "Consummatumest" (It is consummated!) Jesus spoke 7 times on the Cross (Lk 23:34-note; Jn 19:36-note; Lk 23:42-note; Mt 27:46-note; Jn 19:28,30-note, Lk 23:46-note). TETELESTAI was Jesus' next to last word and was followed by "Father, INTO THY HANDS I COMMIT MY SPIRIT." (Lk 23:46-note). "IT IS FINISHED!"

As C. H. Spurgeon says, TETELESTAI conveys "an ocean of meaning in a drop of language, a mere drop. It would need all the other words that ever were spoken, or ever can be spoken, to explain this one word. It is altogether immeasurable. It is high; I cannot attain to it. It is deep; I cannot fathom it. IT IS FINISHED is the most charming note in all of Calvary's music. The fire has passed upon the Lamb. He has borne the whole of the wrath that was due to His people. This is the royal dish of the feast of love."

This did not occur in the garden or the cross. The cross and the garden are places of pure Love, Forgiveness, and Reconciliation. The unbeliever on the cross sentenced himself to damnation because he chose not to believe.

The Garden of Gethsemane and the scene at the cross with the thieves represent the reconciliation of humans to God in its purest form. Jesus demonstrated pure love by asking for the forgiveness of man immediately after a wrong is done by the said man. This is why someone can be forgiven and restored immediately after doing wrong. Note, it is not about the wrongdoing, but it is about the wrong doer being sorry for the wrong, or the forgiver not wanting to carry the baggage of the wrong done. No wonder forgiveness is a benefit to the forgiver and the forgive-ee.

Luke 22:44: "And being in an agony he prayed more earnestly: and his sweat was as it were great drops of blood falling down to the ground."

Jesus' agony was represented by how he sweated. Often, as humans, we sweat as we go through situations of fear, stress, worry, and hardship – sometimes to our lament. Jesus' agony was so great that he literally sweated blood. This was recorded so that you and I could how much he sacrificed for us. Jesus paid an emotional price, in addition to a physical price. The physical price was the crucifixion. The emotional price was the agony and stress that he experienced, which was so great that he sweated "drops of blood". Can you imagine a human literally sweating blood? That would suggest they are proceeding unto death, or death is nigh. For Jesus, death was indeed nigh, and he agonized about it. Though he was in distress, he knew the reward was greater than the agony he endured, so he pushed forward to execute the Father's plan.

Matthew 27:46: "And about the ninth hour Jesus cried with a loud voice, saying, Eli, Eli, lama sabachthani? that is to say, My God, my God, why hast thou forsaken me?"

Now, while on the cross, he went through another rather agonizing experience. This time he bled for real. He was nailed to the cross for our sins, but the real agony came when he realized that God had forsaken him. Jesus, having taken on the sins of the world, could not have avoided this experience, as God in his righteousness cannot tolerate sin. This too is recorded and demonstrated, so that we will understand the depth of Jesus' sacrifice and the reason why the unbelieving man automatically takes unto himself condemnation. God cannot tolerate sin and those who do not choose Christ will likewise be forsaken by God in the end. However, because He cannot have anything to do with sin, even if He tried, He is righteous and just altogether.

Jesus was rejected by God, not the Father. They are two different capacities and authorities. This rejection did not last long, but Hell is everlasting; it will last forever. Therefore, the rejection

Christ suffered on the cross is way less than the rejection man will face if they do not believe.

I witnessed the Damnation of man:

2 Thessalonians 2:7-9: And to you who are troubled rest with us, when the Lord Jesus shall be revealed from heaven with his mighty angels, in flaming fire taking vengeance on them that know not God, and that obey not the gospel of our Lord Jesus Christ: who shall be punished with everlasting destruction from the presence of the Lord, and from the glory of his power;

John 3:19-20: And this is the condemnation, that light is come into the world, and men loved darkness rather than light, because their deeds were evil. For every one that doeth evil hateth the light, neither cometh to the light, lest his deeds should be reproved.

The wisest man in the Bible Solomon reminds me why God seeks an upright man. "This only have I found: God created mankind upright, but they have gone in search of many schemes" (Ecclesiastes 7:29). It will make sense because the Bible says that sin enters when man first desires them.

Good and Perfect Gifts

James 1:14-17: 14But each one is tempted when by his own evil desires, he is lured away and enticed. 15Then after desire has conceived, it gives birth to sin; and sin, when it is fully grown, gives birth to death. 16Do not be deceived, my beloved brothers. 17Every good and perfect gift is from above, coming down from the Father of the heavenly lights, with whom there is no change or shifting shadow.

Death in Adam, Life in Christ

11Not only that, but we also rejoice in God through our Lord Jesus Christ, through whom we have now received reconciliation. 12Therefore, just as sin entered the world through one man, and death through sin, so also death was passed on to all men because all sinned. 13For sin was in the world before the law was given, but sin is not taken into account when there is no law.

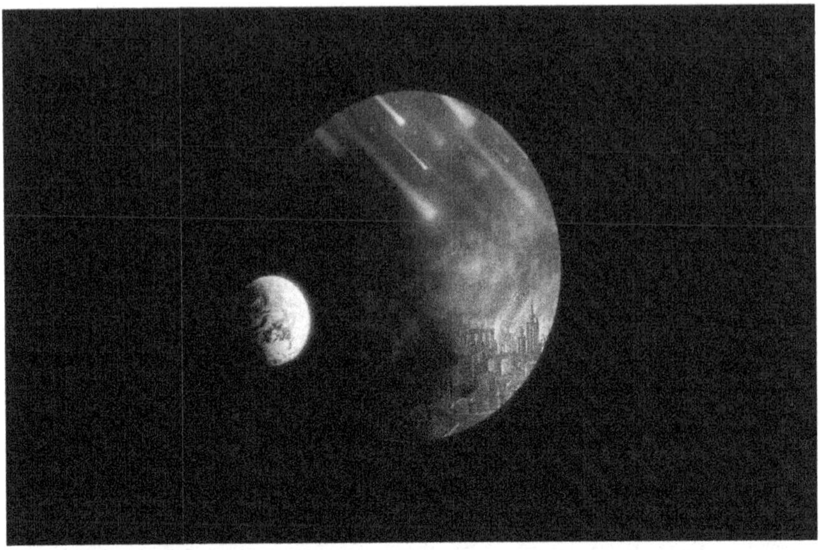

Matthew 25:41: "Then he will say to those on his left, 'Depart from me, you accursed, into the eternal fire that has been prepared for the devil and his angels!'"

I was coming back from the library with my son when I felt a presence that proved difficult for me to discern.

"Are you able to see an angel around us," I asked my son whose prophetic gift was just beginning to develop.

I knew the feeling was strange and new and quite different from my spiritual awareness of God, Jesus Christ, and the Holy ghost. Although, this feeling was different from my anointed atmosphere, however, it didn't feel demonic. I felt a company, but I didn't see it.

My son then said, "Mom, you have a little boy by you. His name, maybe, is Michael or Jacob as I remember."

I asked him what the little boy wanted.

"He wants to go to heaven. Somebody told him he should come to you to take him to heaven if he would follow you", my son replied.

"I do travel to heaven but how did he know that?" I asked him.

He stated that he is a spirit of course, and it must be a very young spirit. So, he needs to follow me to heaven.

"It's ok, but nobody follows me to heaven because God doesn't allow it. It's His secret place, a private place. I don't even know, but anybody that follows me could be arrested. Obviously, I cannot answer that," I reacted.

I wondered why he wanted to go to heaven with me.

"He fell," my son said when I asked where he was coming from. He fell down and he was not coming up. I didn't know what to do so I called one of the pastors. This situation was beyond me; I don't deal with ghosts.

I only had two ghosts I spoke with the day a church member in a church I ministered in died. I spoke to another spirit that was dead. At the time, I didn't know if it was dead or not; he just spoke to me in my dream. I was about to wake him up at a church where I was ministering in Chicago when he said, "Sister Julie, please don't wake me up". Behold, it was in my dream and I won't say I really spoke to him. I just dreamt about him asking me if that was a good place to go. He begged me not to go to his funeral and wake him. I stopped praying for dead to resurrect after that unless asked. I am a warrior and I hate to give Satan even an inch of space to violate God's will in people's lives.

Encounter with heaven and hell.

Anyway, as I'm walking with my son, I learn that there's a little boy who wants to go to heaven with me. I can't even take my own little boy to heaven, if I wanted to, so I knew I couldn't personally take this boy to heaven. On this journey, everybody is on their own. I called a Christian brother, and he didn't know what to do either. He reminded me that when Christ was resurrected, the dead resurrected with him. The Bible didn't quite tell us whether they all resurrected and went to heaven with Him. Maybe they are here or maybe they are somewhere else; we are not sure.

"Ok, forget it," I said. "I keep hearing people say to send him to the light. I don't know how to send him to the light. The only light I know is Jesus".

I bowed my head and I said, "Jesus you are the light of the world and this boy is looking for guidance to go to heaven."

I asked the boy if he knew Jesus.

He replied, "Yes, the big guy up there."

Then I said to Jesus, "he is looking for a home, would you please show him the way to come?".

My son said, "Look, Mom! An escalator, a lighted escalator is coming down!"

He was running to it and a bunch of people were doing so too. Out of the corner of my eye, I saw some white figures. I didn't know what they meant.

"What is going on?" I said to myself.

I asked my son what was going on. I was in my living room waiting. I could tell it was something important because of the light in the corner of my eye. After thirty minutes, nothing happened, and I asked my son if he saw anything ever happen. I waited as I wouldn't go to bed with all those activities going on. I didn't want to have a bad dream. However, my son was busy, and I knew I was disturbing him. I stopped worrying him, bowed and I said a simple prayer: "Father, please open my eyes to see what is going on."

Afterward, I saw someone who looked like a lady with flower clothes standing in line like a mall trying to go to the escalator. It seemed like there was an important place everyone was trying to go and I agreed to it. I didn't know what was going on but I surely wanted to know. After a while, God showed me what was happening. I looked up and there was someone with a book at the gate used for roll call. I said, "Son, they are checking if their names are in the book. They are checking the book before letting people in. I know the Bible says it, but I didn't know I would live to see it." I watched as one person went and another man was sent to the other side.

He was like the pain of a thousand times, as if the pain of this world was on him; he was so sad. I started crying because nobody should feel that pain; it was so bad when his name wasn't in the book. When we reject Jesus Christ here, the damnation is like a removal of life. It feels as if you're continuously holding your breath, while being completely still, but you are still alive. It's like all the pains in the whole world are on your whole body, all the sadness you can imagine. That is how it feels, multiplied by one thousand. This is how it feels to be rejected by Christ: the absence of life, absence of happiness.

He walked me through and gave me permission to follow. I couldn't go any farther because of what I saw. I hadn't seen him go to hell. I told my son the day I went to hell was the most painful experience. God made me more committed to do this mission, committed to pray that nobody goes to hell. I pray that nobody will go to hell and that you will receive Jesus Christ as your Lord and savior. If you add every pain in this world and carry it for a hundred years at least, that is going to evaporate, but this pain is for eternity, forever and ever. Do the calculation and imagine how painful that would be.

None of us will suffer for eternity; none of us will suffer for a hundred years. On earth, we may probably suffer for thirty to forty years but we have some enjoyment mixed with it. There is no enjoyment but a total absence of life in hell, where there is more to go with fire, brimstone and torment. Nobody should go to hell. I remember I used to tease Jesus because of his fanatical love with humanity; I don't know why he loves humanity. The Father came to me and asked why He was disappointed. He told me I was disappointed that a man would make idol smaller than Himself and say, "This is my God".

Sometimes, it may be taller or double- or three-times my size, but compared to God, it's nothing. The very day I stopped teasing Jesus for being a loving fanatic was the day I became the loving fanatic. I prayed the prayer of Jesus Christ on the cross that day. I cried to the Father for the first time from my heart and I said:

"Father, forgive them; they do not know what they are doing. Forgive this generation. Forgive us, because if they had seen what I have seen - the Messiah, if they had seen a loving Father, if they had seen the One who came down for them, if they could seen how tiny they are, if they could see that they are like an ant under a giant, if they could see, they would understand and if they had the capacity to understand, they would not do what they are doing. But they don't know. Therefore, Father, please forgive them for they do not know what they are doing. And I plead with you, please let them hear what the spirit is saying to the church."

I beseech you to embrace the Lord and believe my words, as I have nothing to gain for your salvation. I don't have a church (I know I may have one in the future). I don't have a ministry yet. I have just been crucified, waiting for the rest of the gospel to be fulfilled. I know I am here in the physical measurement of things as someone God loves.

This life is not worth it. Take away everything in this life and give me God. I will always love Him. Everything on earth is going to melt and all that will save you is your faith in Christ. Believe that God made you from the sand and gave you His breath of life. We are made from the dust and when we die physically, we shall be cremated to dust. The Bible is true. Ask Jesus to come into your life, and tell your neighbors it's not worth it to live without Him.

Chapter 12

ASSURING YOUR SALVATION TO OTHERS

The earth is changing and a lot of people find themselves unable to go to church during the pandemic. I want to attach this foundational guide for any new believers. Whether you are able to go to church or not, whether you are persecuted or not, whether you go to an underground church or worship alone, this guide below helped me. Know that God is everywhere you are. You can worship with family, but it is very helpful to gather in community.

This is also a tool for evangelism, in case you encountered God reading this book in your car or home. It will tell you what is next so you can enjoy unlimited blessings and favor. Whenever you see a question, try to come up with the answer in your mind before reading the answer. This will help you meditate. Before you read this section, please invite the Holy Spirit in prayer to help you. Pray, "Dear Holy Spirit, you are welcome here. Please be my teacher and guide me to eat and retain and grow in the knowledge of God. Please open my eyes to behold thy wondrous works in God's word, in Jesus name, Amen." It is written that if you confess him before men, God will confess you in heaven.

The only way to successfully overcome the enemy is by your confession.

Revelation 12:11: "And they overcame him by the blood of the Lamb, and by the word of their testimony; and they loved not their lives unto the death."

Becoming Like Christ

How do you become a Christian?

Answer: Human beings are, by nature, sinners. We are born into this world as sinners; therefore, it is essential for every human being who is knowledgeable about Jesus Christ to commit, believe, and surrender their life to Jesus Christ. There is no short cut to becoming a follower of Christ except through salvation. Salvation comes from accepting Jesus Christ as your personal savior.

How do you accept Jesus Christ as your savior?

- **Acknowledge only Jesus Christ as your savior**

According to Acts 4:12 (NIV), salvation is found in no one else, for there is no other name under the heaven given to mankind by which we must be saved. In Acts 16:13 (NIV), they replied, "Believe in the Lord Jesus, and you will be saved –you, and your household."

Why Confess and repent your sins?

According to 1 Corinthians 6:9-10 (NIV), "or you do not know that, wrong doers will not inherit the kingdom of God? Do not be deceived: neither the sexually immoral nor idolaters nor adulterers nor men who have sex with men nor thieves nor the greedy nor drunkards nor slanderers nor swindlers will inherit the kingdom of God."

Romans 10:10: for a man believes with his heart and so is justified, and he confesses with his lips and so is saved.

- **Commit to resist sins**

No one can stop sinning by their strength. Thus, Paul said after you are born again, the war of who you are starts. You belong to God, but your body is still accustomed to some behaviors you must no longer do to gain the benefit of doing your righteous desires. When you receive Jesus, you become the righteousness of God. That does not change unless you change your mind. You made a choice to choose Him and only you can leave Him because He is constant and does not change.

Newton's First Law of Motion states that **a body at rest** will remain at **rest** unless an outside force acts on it, and **a body** in motion at a constant velocity will remain. Apply the outside force of your faith and keep going forward. God is not evil; He knows the difference of your weakness and your desire. Keep repeating "I am the righteousness of God" till you become it in your eyes because in His eyes, that is what you will always be. The work God wants from us is to believe. You have done it. You are His.

Titus 2:11-12 (NIV): For the grace of God that offers salvation has appeared to all people, it teaches us to say "No" to ungodliness and worldly passions, and to live self-controlled, upright, and Godly lives in this present age.

Romans 7:19: For I do not do the good I want to do, but the evil I do not want to do--this I keep on doing.

I am the righteousness of God in Christ Jesus! (2 Corinthians 5:21).

Believing in God

1 Peter 1:8-9: Though you have not seen him, you love him; and even though you don't see him now, you believe in him and are filled with an inexpressible, and glorious joy, for you are receiving the end result of your faith, the salvation of your souls.

In John 3:16 (RSV), we are told that God so loved the world that he gave his only son, that whoever believes in Him should not perish but have eternal life. In John 6:29, Jesus said, "The work of God is this: to believe in the one he has sent."

Receive Christ as your savior

John 1:11-13: He came to his home, and His own people received Him not. But to all whom received Him, who believed in His name, he gave power to become children of God: who were born, neither of blood nor of the will of man, but of God.

The Three Aspects of Salvation

Justification: the action of declaring or making righteous in the sight of God

Re-generation: to have a new life in Christ.

Sanctification: being cleansed to be born again in Christ.

1. Bible Study

Salvation is a life-long process. To understand what is expected of you once you give your life to Christ, you need to regularly read the Bible. The Bible is the manual that you need to follow to live the way God wants you to. You are known as Christian at the point of action.

The essential of your Bible study:

- **Ensure you have a Bible**
- **Ensure you have a notebook.**
- **Plan out a designated time to spend before God.**

How to study your Bible effectively?

- **Get your mind ready to hear from God.**
- **Listen to the voice of God.**
- **Understand whether there is any promise to claim from your study.**
- **Understand whether there is any example to follow.**
- **Understand whether there is any error to avoid.**
- **Understand whether there is any divine knowledge to acquire.**

Reasons why Bible studying is important in the life of a Christian

To learn how to fear God

Deuteronomy 31:12-13: "Assemble the people, men, women, and little ones and the sojourner within your towns that they may hear, and learn to fear the LORD your God, and be careful to do all words of the law, and that all children, who have not known it, may hear, and learn to fear the LORD your God, as long you live in the land which you are going over the Jordan to possess."

To become successful in Christian life

Joshua 1:8: "This book of the law shall not depart out of your mouth, but you shall meditate it day and night, that you may be careful to do according to all that is written in it; for then you shall make your way prosperous, and then you shall have good success."

To learn God's mind concerning you

Daniel 9:12: "He has confirmed his words, which he spoke against us and against our rulers who ruled us; by bringing upon us a great calamity, for under heaven there has not been done like of what has been done against Jerusalem."

To get God's approval and grow in God spiritually

2 Timothy 2:15: "Do your best to present yourself to God as one approved, a workman who has no need to be ashamed, rightly handling the word of truth."

1. Fellowship

It is your duty as a believer to choose a proper fellowship for yourself. A proper fellowship is a gathering of two or more people in God's name. In a proper fellowship, brethren gather to know God, walk with Him, and serve Him.

What are the features of a proper relationship of brethrens in Christ?

Working in unity

Romans 12:3-5 states, "For by the grace given to me I bid every one among you not to think himself more highly than he ought to think, but to think with sober judgment, each according to the measure of faith which God has assigned him; for as in one body we have many members, and all the members do not have the same function, so we though many, are one body of Christ."

1 Corinthians 12:12-13 emphasizes, "for just as the body is one and has many members, and all the members of the body, though many are one body, so it is with Christ, for by one Spirit we were all baptized into one body; Jews, Greeks, slaves or free –all were made to drink of one spirit." Ephesians 4:4 emphasizes the same. In addition, Psalms 133:1-2 says, "Behold, how good and pleasant it is when brothers dwell in unity! It is like the precious oil upon the head, running down upon the beard of Aaron, running down the collar of his robes."

3. Holy Spirit and Water Baptism

A believer without the Holy Spirit is dead. It is the will of God to fill men with the power of the Holy Spirit. Anyone in flesh who believes in Christ qualifies to be filled with the power of the Holy Spirit. According to Luke 24:49, the Holy Spirit is the promise of God the father to His children. The verse states, "And behold, I send the promise of my father upon you; but stay in the city until you are clothed with the power from the most high."

Water baptism is also important in your walk with Jesus Christ as a believer. The Bible talks about water immersion in which a believer makes a public confession of their faith.

What is the process needed to receive the Holy Spirit?

- **Must be Saved**

Acts 10:43 says, "to him all the prophets bear witness that every man who believes in him receives forgiveness of sins through his name." While Peter was still saying this, the Holy Spirit fell on all who heard the Word.

Acts 9:17: "So Ananias has departed and entered the house. And laying his hands on him he said, 'Brother Saul, the Lord Jesus who appeared to you on the road by which came, has sent me that you may regain your sight and be filled with the Holy Spirit.'"

- **Understanding/Knowing about the Holy Spirit**

Understanding the Holy Spirit is the key to receiving Him. Understanding who the Holy Spirit is helps us to value Him. As said in John 7:38, "he who believes in me as the scripture has said, out of his heart shall flow rivers of the living water."

- **Pray for the Forgiveness of your Sins**

According to John 14:15-17, the Holy Spirit has a way in a clean soul. It says "If you love me, you will keep my commandments. And I will pray the Father, and He will give you another counselor, to be with you forever, even the Spirit of truth, whom the world cannot receive because it neither sees Him nor knows Him, for He dwells with you and will be in you."

2 Timothy 2:20-21: "In a great house there are not only vessels of gold and silver but also of wood and earthenware, and some for noble use, some for ignoble. If anyone purifies himself from what is ignoble, then he will be a vessel of noble use, consecrated and useful to the master of the house."

Also read Acts 8:18.

- **Desire to be Filled**

1 Corinthians 14:1: "Follow the way of love and eagerly desire gifts of the Spirit, especially prophecy. For anyone who speaks in a tongue does not speak to people but to God."

- **Step out in Faith**

Acts 2:1: "When the day of Pentecost had come, they were all together in one place."

Pray to be Filled

Luke 11:13: "If you then who are evil, know how to give good gifts to your children, how much will the heavenly father give the Holy Spirit to them who ask Him!"

Why does a believer need baptism in the Holy Spirit?

- **To worship God in manner He requires**

According to John 4:24, the Holy Spirit helps us to worship God properly. "God is spirit, and those who worship him must worship in spirit and truth."

- **To teach him how to pray**

Baptism of the Holy Spirit releases the power of prayer.

Romans 8:26: "In the same way, the Spirit helps us in our weakness. We do not know what we ought to pray for, but the Spirit himself intercedes for us through wordless groans."

- **Understanding of God's Word**

The Holy Spirit gives us an understanding of God's Word. Acts 8:29-35 says, "[26]Now an angel of the Lord said to Philip, 'Go south to the road—the desert road—that goes down from Jerusalem to Gaza.' [27]So he started out, and on his way he met an Ethiopian eunuch, an important official in charge of all the treasury of the Kandake (which means "queen of the Ethiopians"). This man had gone to Jerusalem to worship, [28]and on his way home was sitting in his chariot reading the Book of Isaiah the prophet. [29]The Spirit told Philip, 'Go to that chariot and stay near it.'

[30]Then Philip ran up to the chariot and heard the man reading Isaiah the prophet. 'Do you understand what you are reading?' Philip asked.

[31]"How can I," he said, "unless someone explains it to me?" So he invited Philip to come up and sit with him.

[32]This is the passage of Scripture the eunuch was reading:

"He was led like a sheep to the slaughter,

and as a lamb before its shearer is silent,

so he did not open his mouth.

[33]In his humiliation he was deprived of justice.

Who can speak of his descendants?

For his life was taken from the earth."

[34]The eunuch asked Philip, 'Tell me, please, who is the prophet talking about, himself or someone else?' [35]Then Philip began with that very passage of Scripture and told him the good news about Jesus."

- **Rest and refreshment**

Isaiah 28:11-12: "Nay but by men of strange lips and with an alien tongue the Lord will speak to his people, to whom he has said, 'This is rest; give rest to the weary; and this is repose', yet they would not hear.

Evangelism

The primary reason to be filled with the power of the Holy Spirit is to receive salvation. The climax of the Holy Spirit is to become productive. Productivity means you are influencing souls of men into God's kingdom. This art of winning souls to Christ is what we call evangelism.

Why should believers involve themselves in Evangelism?

- **It's a commandment from God**

According to Matthew 28:18, "Jesus came and said to them, 'All authority in heaven and on earth has been given to me. Go therefore and make disciples of all nations, baptizing them in the name of the father, and of the son and of the Holy Spirit, teaching them to observe all that I have commanded you.'" Therefore, evangelism is a mandate for all who are washed by the blood of Jesus Christ.

- **It's an opportunity to win/save souls for Christ**

In the gospel of Mark 16:15, Jesus said "Go into the entire world and preach the gospel to the whole creation. He who believes and is baptized will be saved; but he who does not will be condemned." Romans 1:15-16 says, "So I am eager to preach the gospel to you also who are in Rome. For I am not ashamed of the gospel: it is the power of God for salvation to everyone who has faith, to the Jew first and also to the Greek."

- **A way of giving glory to God**

2 Corinthians 5:18-20 says, "with him we are sending the brother who is famous for preaching of the gospel; and not only that, but he has been appointed by the churches to travel with us in this gracious work which we are carrying on, for the glory of the Lord and to show our good will."

- **Sign of obedience to God**

Acts 5:28-29: "And the high priest questioned them, saying, 'We strictly charged you not to teach in this name, yet here you have filled Jerusalem with your teaching and you intend to bring this man's blood upon us.' But Peter and the apostles answered, 'We must obey God rather than men.'"

- **As testimony to the world**

Mathew 24:14 says, "This gospel of the kingdom will be preached throughout the whole world, as a testimony to all nations; and then the end will come."

1. Praise and Worship

Worship is a very crucial part of being a follower of Jesus Christ. Any time you worship, you express your thankfulness and awe to the Lord.

Praise and worship begin at the individual level. They should take you into the presence of the Lord. God accepts praise and worship depending on your personal motive. For praise and worship to be effective before God, it should involve the whole you: the body, soul and the spirit.

How can you demonstrate your praise and worship to God?

- **Through giving your property to God**

According to Proverbs 3:9-10, you should "honor the Lord with your substances and with the **first fruits** of all your produce; then your bans will be filled with plenty."

Malachi 3:10 talks about giving our tithe. "Bring **full tithes** into the storehouse, that there may be food in my house; and thereby put me into test, says the lord of hosts, if I will not open the windows of heaven for you and pour for you an overflowing blessing." You also worship God through your **offerings.** 1

Chronicles 16:29 says, "Ascribe to the Lord the glory due to his name; bring an offering, and come before him! Worship the Lord in holy array."

- **Through giving your life**

According to Matthew 10:38-39, "and he who does not take his cross and follow me is not of me. He who finds his life will lose it, and he who loses his life for my sake will find it." Likewise, Romans 12:1 says, "I appeal to you therefore, brethren, by the mercies of the God, to present your bodies as living sacrifices, holy, and acceptable to God, which is your spiritual worship."

- **Through your songs**

According to Psalms 100:1, a believer should make a joyful noise to the Lord, all the lands! Serve the lord with gladness! Come into his presence singing.

2. Prayer

If you are looking for the power that can move mountains, then I will tell you that prayer is the greatest force on earth. The most effective prayer is personal prayer. Therefore, as a believer, you must sacrifice your time for prayer. God is always available to us depending on the level of our sacrifice. Prayer is a spiritual way of communication to God.

Key secrets to effective prayer

- **Pray through the Holy Spirit**

The Holy Spirit carries the voice of God into your life. The Holy Spirit dwells in you all the time if you are saved and free from sins. When you are praying, be humble, and surrender to the Holy Spirt so that He can teach you how to pray.

According to Romans 8:26, "likewise the Spirit helps us in our weaknesses; for we do not know how to pray as we ought, but the Spirit Himself intercedes for us with sighs too deep for words."

- **Build personal relationship with God**

Your effectiveness in prayer highly depends on your obedience to the word of God. John 15:7 says, "If you abide in me, and my words abide in you, ask whatever you will, and it shall be done for you."

- **Live a righteous life**

According to Proverbs 15:29, "The Lord is far from the wicked, but He hears the prayers of the righteous." Also, in 1 Peter 3: 11–12, "let him turn away from evil and do right; let him seek peace and pursue it, for the eyes of the Lord are upon the righteous."

State your case before God

In **Exodus 32:11,** Moses brings his request to God in prayer: **"**But Moses besought the Lord his God and said, 'O Lord, why does thy wrath burn hot against thy people, whom thou hast brought out of the land of Egypt with a great power and mighty hand?'"

James 4:2 says, "You desire and do not have; so you kill. You covet and cannot obtain; so you fight and wage war. You do not have because you didn't ask."

Prayer Request:

Chapter 13

Comfort for God's People

"And our hope for you is steadfast, because we know that as you are partakers of the sufferings, so also you will partake of the consolation." – 2 Corinthians 1:7

God's heart is really good, and His plans for His people are likewise good. Sounds untrue, right? You may wonder in your heart: "if God was really good, then why does He sometimes allow unpleasant things to happen to people?" Well, the truth is that God respects people's choices and 'allows' things to happen to them based on their choice of sin, disobedience or stubbornness of heart.

Walking with God is the most adventurous journey any man could have in this life; this is because we often meet a lot of surprises on the way, some good and some bad. Take for example when the Lord brought Israel out of Egyptian slavery. Instead of taking them straight to their promised land through a nearby road, which would only take a few days walk, the Lord chose to lead them through a much longer road.

Exodus 13:17-18: "Then it came to pass, when Pharaoh had let the people go, that God did not lead them by way of the land of the Philistines, although that was near; for God said, 'Lest perhaps the people change their minds when they see war and return to

Egypt.' So God led the people around by way of the wilderness of the Red Sea. And the children of Israel went up in orderly ranks out of the land of Egypt."

The Lord wanted total liberation for His people, knowing that the Israelites had been builders (and not soldiers) for the past 400 years. If He sent them on the easy road, they would have met war unprepared and it would have forced them to return to Egypt.

Most of God's people all over the world usually had a similar experience when we first surrendered our lives to the Lord. Things seemed to go wild, and the devil often tried to make us think we made a terrible mistake getting born again. The devil manipulates a series of attacks against us in a bid to make us turn back to 'Egypt'. Whenever God allows the devil to do unpleasant things to His own people, it is so He could demonstrate His wisdom and power in our lives.

But here's the good news! God's ability to deliver, provide and protect is far greater than the devil's tricks to weaken our faith and cut off our walk with God. So, cheer up, saints! The Son of God is your strong Anchor and He will not allow the devil snatch you out of His mighty hands! (See John 10:28-30, Hebrews 6:19)

Are you troubled in any way? Are you challenged by life's unpleasant situations? Your God is mighty to save, heal and deliver! He is coming for you, and He will not delay much longer. However, keep your eyes on the Lord and His glorious power; and not on the devil's tricks and illusions **(See Matthew 14:25-31).**

One way to keep your faith strong and to experience victory after victory is to be full of praises to the Lord (See Philippians 3:1, Habakkuk 3:17-18). It keeps God's presence around you and activates the mighty hand of God to fight your battles. Jesus is able to save and to deliver as you keep your faith in His glorious power!

Hebrews 7:25: "Therefore he is also able to save to the uttermost those who come to God through Him, since He always lives to make intercession for them."

Credit for Science & Bible Converge

https://biblehub.com/acts/1-6.htm

https://biblehub.com/acts/1-7.htm

https://www.kingjamesbibleonline.org/Jeremiah-29-11/

https://www.biblestudytools.com/isaiah/64-4.html

http://www.learnthebible.org/sermon-outlines/the-wonderful-power-of-music.html

https://www.biblegateway.com/passage/?search=Job+38%3A3&version=KJV

https://www.kingjamesbibleonline.org/Job-38-4/

https://www.kingjamesbibleonline.org/Job-38-5/

https://www.kingjamesbibleonline.org/Job-38-7/

https://bibleapps.com/genesis/1-2.htm

https://biblehub.com/psalms/104-4.htm

https://biblehub.com/m/hebrews/1-7.htm

https://biblehub.com/genesis/6-1.htm

https://biblehub.com/galatians/5-16.htm

https://www.youtube.com/watch?v=RGOQLw0DsIQ

https://www.bible.com/bible/59/2PE.2.4.ESV

https://connectusfund.org/luke-10-19-meaning-of-verse-with-simple-commentary

Credit for Science & Bible Converge

https://biblehub.com/isaiah/14-12.htm

https://biblehub.com/matthew/7-6.htm

https://biblehub.com/matthew/7-7.htm

https://biblehub.com/matthew/7-8.htm

https://www.thedailybibleverse.org/bible-verse-early-in-the-morning-they-left-for-the-desert-of-tekoa-as- they-set-out-jehoshaphat-stood-and-said

https://biblehub.com/2_chronicles/20-20.htm

https://www.biblegateway.com/verse/en/2%20Chronicles%2020%3A20

http://foodforthoughtministry.blogspot.com/2009_06_20_archive.html

https://biblehub.com/matthew/18-19.htm

https://biblehub.com/matthew/21-22.htm

https://biblehub.com/mark/11-24.htm

https://biblehub.com/jeremiah/29-11.htm

https://biblehub.com/kjv/isaiah/55-9.htm

https://www.youtube.com/watch?v=5bWHSpmXEJs

https://biblehub.com/matthew/5-14.htm

https://www.kingjamesbibleonline.org/Matthew-5-15/

https://www.biblegateway.com/passage/?search=Matthew+5%3A16&version=KJV

Credit for Science & Bible Converge

https://biblehub.com/acts/10-38.htm

http://shamah-elim.info/hivite.htm

https://www.youtube.com/watch?v=xuKL759G9sk

https://www.kingjamesbibleonline.org/Matthew-21-16/

https://www.bible.com/bible/1/MAT.21.16.KJV

https://znanija.com/task/24281868

https://m.facebook.com/story.php?story_fbid=267649953925571&id=496928613721123

https://www.bible.com/bible/111/HEB.2.3-4.NIV

https://biblehub.com/hebrews/2-3.htm

https://biblehub.com/hebrews/2-4.htm

https://www.kingjamesbibleonline.org/1-Thessalonians-1-5/

https://biblehub.com/commentaries/romans/8-38.htm

https://biblehub.com/2_corinthians/13-14.htm

https://www.youtube.com/watch?v=dkK_fE2en-Q

https://bible.knowing-jesus.com/John/15/19

https://www.biblegateway.com/passage/?search=John+15%3A18-20&version=NKJV

https://www.biblegateway.com/passage/?search=John+15%3A20&version=ESV

https://biblehub.com/john/15-20.htm

https://www.biblestudytools.com/bible-study/explore-the-bible/3-surprising-ways-i-can-do-all-things- through-christ-who-strengthens-me.html

https://biblehub.com/2_corinthians/11-24.htm

https://www.biblestudytools.com/2-corinthians/11-25.html

https://www.biblica.com/bible/niv/2-corinthians/11/

https://biblehub.com/2_corinthians/11-26.htm

https://www.biblegateway.com/passage/?search=2+Corinthians+11%3A27&version=NIV

https://www.bible.com/bible/111/2CO.11.28.NIV

https://biblehub.com/2_corinthians/11-29.htm

https://www.biblegateway.com/passage/?search=2+Corinthians+11%3A29&version=NIV

http://watchmanchow.blogspot.com/2011/04/colossians-323-says-whatever-you-do.html

https://biblehub.com/isaiah/6-8.htm

https://thesoftpinktruth.bandcamp.com/album/shall-we-go-on-sinning-so-that-grace-may-increase

https://biblehub.com/romans/6-2.htm

https://www.biblestudytools.com/romans/6-3.html

https://biblehub.com/matthew/5-44.htm

https://biblehub.com/matthew/5-45.htm

https://biblehub.com/philippians/2-15.htm

https://biblehub.com/1_corinthians/10-13.htm

https://www.kingjamesbibleonline.org/1-Corinthians-10-13/

https://biblehub.com/m/1_peter/4-14.htm

https://www.christianity.com/bible/bible.php?q=1+Corinthians+4%3A10&ver=esv

https://biblehub.com/1_corinthians/4-11.htm

https://biblehub.com/1_corinthians/4-12.htm

https://biblehub.com/1_corinthians/4-13.htm

Credit for Science & Bible Converge

https://biblehub.com/1_corinthians/4-14.htm

https://www.youtube.com/watch?v=JLs-YPJCksk

https://www.biblestudytools.com/matthew/2-2.html

https://biblehub.com/matthew/2-2.htm

https://biblehub.com/par/matthew/2-9.htm

https://www.slideshare.net/faithworksc/stand-3-stand-strong-ptr-alvin-gutierrez-10am-morning-service- 67231824

https://biblehub.com/romans/8-29.htm

https://biblehub.com/romans/8-30.htm

https://www.biblestudytools.com/topical-verses/inspirational-bible-verses/

https://biblehub.com/romans/8-31.htm

https://www.simplybible.com/f162.htm

https://www.biblegateway.com/passage/?search=Hebrews+9%3A27&version=KJV

https://biblehub.com/hebrews/12-1.htm

https://www.ted.com/talks/simon_sinek_how_great_leaders_inspire_action/transcript

https://biblehub.com/revelation/1-7.htm

https://www.kingjamesbibleonline.org/Hebrews-12-22/

https://www.ucg.org/bible-study-tools/booklets/angels-gods-messengers-and-spirit-army/the-vast- multitudes-of-heavens-armies

https://biblehub.com/hebrews/12-24.htm

Credit for Science & Bible Converge

https://www.kingjamesbibleonline.org/Hebrews-12-25/

https://www.kingjamesbibleonline.org/Hebrews-12-26/

https://www.kingjamesbibleonline.org/Hebrews-12-27/

https://www.kingjamesbibleonline.org/Hebrews-12-28/

https://bible.knowing-jesus.com/Romans/11/28

https://biblehub.com/romans/11-29.htm

https://bible.knowing-jesus.com/Romans/11/30

https://bible.knowing-jesus.com/Romans/11/31

https://bible.knowing-jesus.com/Romans/11/32

https://www.biblegateway.com/passage/?search=Galatians%205%3A16-18&version=ESV

https://www.desiringgod.org/messages/the-war-within-flesh-versus-spirit

https://biblehub.com/galatians/5-18.htm

https://www.biblegateway.com/verse/en/Matthew%2010%3A27

https://biblehub.com/matthew/10-28.htm

https://biblehub.com/matthew/10-29.htm

https://www.cdc.gov/nchs/fastats/deaths.htm

https://biblehub.com/matthew/10-28.htm

https://biblehub.com/hebrews/10-31.htm

https://www.biblegateway.com/passage/?search=2+Samuel+12&version=NCV

https://www.biblestudytools.com/ncv/2-samuel/12-4.html

https://bible.knowing-jesus.com/2-Samuel/12/5

https://biblehub.com/2_samuel/12-7.htm

https://www.biblegateway.com/verse/en/2%20Samuel%20 12%3A8

https://www.biblestudytools.com/ncv/2-samuel/12-9.html

https://www.biblestudytools.com/ncv/2-samuel/passage/?q=2-samuel+12:10-12

https://www.crunchyroll.com/forumtopic-914941/how-do-you-feel-about-abortion?pg=45

https://www.bible.com/bible/105/2SA.12.17.NCV

https://znanija.com/task/23060628

https://sharonserrano8.wordpress.com/2017/11/29/doubting-towards-faith/

https://www.bible.com/bible/105/2SA.12.20.NCV

https://www.biblestudytools.com/ncv/2-samuel/12-21.html

http://www.justblog.me/2013/11/happy-6th-birthday-boys.html

https://www.nairaland.com/4407941/nysc-hail-thee-lady-married

https://biblehub.com/galatians/6-7.htm

https://biblehub.com/galatians/6-8.htm

https://biblehub.com/job/4-8.htm

https://biblehub.com/job/13-9.htm

https://www.biblestudytools.com/acts/13-22.html

https://biblehub.com/acts/13-22.htm

https://biblehub.com/genesis/15-4.htm

https://biblehub.com/genesis/21-17.htm

https://biblehub.com/genesis/21-18.htm

https://biblehub.com/2_samuel/12-13.htm

https://biblehub.com/2_corinthians/5-20.htm

https://biblehub.com/2_corinthians/5-21.htm

https://www.biblegateway.com/passage/?search=John+20

&version=NIV

https://julieyipwilliams.wordpress.com/2017/10/16/complete/

https://www.youtube.com/watch?v=MMXPnSxEKL0

https://www.billkochman.com/VerseLists/verse193.html

https://biblehub.com/hebrews/12-3.htm

https://biblehub.com/hebrews/12-2.htm

https://www.linkedin.com/pulse/why-im-taking-5-weeks-out-summer-mark

https://answers.yahoo.com/question/index?qid=20091107104453AAURpIl

http://www.kingjamesbibles.com/proverbs_21.html

https://biblehub.com/genesis/15-5.htm

https://www.bible.com/bible/111/GEN.1.16.NIV

https://biblehub.com/par/psalms/147-4.htm

https://www.biblestudytools.com/dictionaries/bakers-evangelical-dictionary/new-heavens-and-a-new- earth.html

https://biblehub.com/matthew/6-20.htm

https://biblehub.com/revelation/3-20.htm

https://biblehub.com/revelation/3-21.htm

https://bible.knowing-jesus.com/Revelation/3/22

https://biblehub.com/romans/8-34.htm

https://www.kingjamesbibleonline.org/Romans-8-35/

https://biblehub.com/romans/8-35.htm

https://biblehub.com/1_corinthians/15-8.htm

https://biblehub.com/1_corinthians/15-9.htm

https://www.biblestudytools.com/revelation/2-28.html

https://www.kingjamesbibleonline.org/Genesis-17-1/

https://biblehub.com/genesis/17-7.htm

Credit for Science & Bible Converge

https://www.kingjamesbibleonline.org/Genesis-17-8/

https://www.kingjamesbibleonline.org/Genesis-17-9/

https://www.nasa.gov/mission_pages/apollo/apollo11_audio.html

https://www.nasa.gov/news/debris_faq.html

https://answers.yahoo.com/question/index?qid=20100501150715AA5NUlZ

https://answers.yahoo.com/question/index?qid=20160120114432AACEEpY

http://thingsworthknowin.blogspot.com/2016/03/how-many-solar-systems-are-in-our-galaxy.html

https://spaceplace.nasa.gov/review/dr-marc-space/solar-systems-in-galaxy.html

https://www.thesoloreads.com/2019/02/how-galaxies-are-formed.html

http://thingsworthknowin.blogspot.com/2016/03/how-many-solar-systems-are-in-our-galaxy.html

https://answers.yahoo.com/question/index?qid=20160120114432AACEEpY

https://answers.yahoo.com/question/index?qid=20100501150715AA5NUlZ

https://imgur.com/gallery/GtTJp7p

https://answers.yahoo.com/question/index?qid=20080713053439AA4U3GE

https://biblehub.com/psalms/19-1.htm

https://web.mit.edu/jywang/www/cef/Bible/NIV/NIV_Bible/PS+19.html

https://biblehub.com/isaiah/65-17.htm

Credit for Science & Bible Converge

https://www.biblegateway.com/passage/?search=2+Peter+3%3A13&version=NIV

https://biblehub.com/numbers/24-17.htm

https://www.prayertents.com/b?search=Numbers+24:17&version=6

https://biblehub.com/kjv/romans/6-16.htm

https://bible.knowing-jesus.com/Isaiah/64/5

https://bible.knowing-jesus.com/Isaiah/64/6

https://biblehub.com/isaiah/64-6.htm

https://bible.knowing-jesus.com/Isaiah/64/7

https://biblehub.com/john/1-1.htm

https://www.kingjamesbibleonline.org/John-1-2/

https://www.biblestudytools.com/tmba/john/1-3.html

https://www.kingjamesbibleonline.org/John-1-4/

https://www.kingjamesbibleonline.org/John-1-5/

https://www.biblegateway.com/passage/?search=John+1&version=KJV

https://www.kingjamesbibleonline.org/John-1-7/

https://www.kingjamesbibleonline.org/John-1-9/

https://www.kingjamesbibleonline.org/John-1-8/

Credit for Science & Bible Converge

https://biblehub.com/john/1-10.htm

https://www.kingjamesbibleonline.org/John-1-11/

https://www.kingjamesbibleonline.org/John-1-12/

https://www.kingjamesbibleonline.org/John-1-14/

https://www.biblegateway.com/passage/?search=John+1&version=KJV

https://www.kingjamesbibleonline.org/John-1-16/

https://www.kingjamesbibleonline.org/John-1-18/

https://www.kingjamesbibleonline.org/John-1-17/

https://medium.com/publishous/what-excuses-do-you-have-not-to-write-f7e3326baee0

https://blog.logos.com/2017/11/god-wasnt-alone-created-world-says-bible/

https://www.biblestudytools.com/leb/job/38-4.html

https://bnonn.com/what-is-the-kingdom-of-god-3/

https://www.puritanboard.com/threads/the-nephilim.91956/page-2

https://daylidigestnews.wordpress.com/

https://www.scribd.com/book/365096077/The-Unseen-Realm-Recovering-the-Supernatural-Worldview-of- the-Bible

https://www.bible.com/bible/59/MAT.2.1-2.ESV

https://biblehub.com/matthew/2-2.htm

Credit for Science & Bible Converge

https://www.bible.com/bible/59/NUM.24.17.ESV

https://biblehub.com/numbers/24-17.htm

https://philressler.com/verse-of-the-week-matthew-211/

https://www.crosswalk.com/special-coverage/christmas-and-advent/gold-frankincense-and-myrrh.html

https://blog.logos.com/2017/11/god-wasnt-alone-created-world-says-bible/

https://www.sciencealert.com/scientists-just-captured-the-actual-flash-of-light-that-sparks-when-sperm- meets-an-egg

https://birthfaith.org/mothering/monday-miscellany

https://www.youtube.com/watch?v=efVUMPdZcPo

https://www.nairaland.com/3072223/first-time-scientist-discovers-exactly

https://www.sott.net/article/346010-Researchers-discover-a-flash-of-light-that-occurs-at-the-moment-of- conception

https://elemental.medium.com/the-problem-with-supplements-e0f5f2fb1d74

https://translate.yandex.ru/

https://www.dailymail.co.uk/sciencetech/article-3559563/Sparks-really-fly-new-life-formed-Images-reveal-fireworks-occur-moment-sperm-meets-egg.html

https://www.sciencealert.com/scientists-just-captured-the-actual-flash-of-light-that-sparks-when-sperm- meets-an-egg

https://cns.utexas.edu/news/when-sperm-meets-egg-zinc-fireworks-on-display

https://translate.yandex.ru/

https://www.lifesitenews.com/opinion/light-at-conception-gods-creation-not-ours

https://www.christianheadlines.com/columnists/breakpoint/new-discovery-of-light-at-conception-should- direct-focus-to-god-s-creation-not-ours.html

https://www.youtube.com/watch?v=dZezp1pyRao

https://www.compellingtruth.org/names-Holy-Spirit.html

https://www.puritanboard.com/threads/the-nephilim.91956/page-2

https://greatnessforhumanity.com/659-2/

https://biblehub.com/genesis/6-2.htm

https://www.bible.com/bible/111/GEN.2.7.NIV

https://www.bible.com/bible/111/JOB.33.4.NIV

http://topverses.com/Bible/Job/27/3

https://www.biblestudytools.com/john/20-22.html

https://biblehub.com/ezekiel/37-9.htm

http://jaymieloveschrist.blogspot.com/2008/10/ephesians.html

https://www.biblestudytools.com/ephesians/1-4.html

https://www.biblegateway.com/passage/?search=eph+1%3A3-6&version=NIV

https://www.youtube.com/watch?v=2nBA-j6ot0I

https://biblehub.com/m/job/27-3.htm

https://www.facebook.com/Harmonyof1/posts

https://bible.knowing-jesus.com/John/6/63

https://biblehub.com/john/6-63.htm

https://biblehub.com/hebrews/12-29.htm

https://biblehub.com/deuteronomy/9-3.htm

https://biblehub.com/par/psalms/104-4.htm

https://biblehub.com/matthew/5-14.htm

https://www.kingjamesbibleonline.org/Matthew-5-15/

https://www.biblegateway.com/passage/?search=Matthew+5%3A16&version=KJV

Credit for Science & Bible Converge

https://biblehub.com/isaiah/63-10.htm

https://www.biblestudytools.com/matthew/12-31-compare.html

https://biblehub.com/ephesians/4-30.htm

https://biblehub.com/hebrews/10-29.htm

https://www.crosswalk.com/faith/bible-study/who-is-the-holy-spirit-5-things-you-need-to-know.html

https://biblehub.com/nehemiah/9-20.htm

https://www.youtube.com/watch?v=rtqTNsc8UMM

https://biblehub.com/ephesians/1-4.htm

https://www.youtube.com/watch?v=wk2y4eu37_s

https://www.gotquestions.org/personal-Savior.html

https://biblehub.com/revelation/21-4.htm

https://biblehub.com/revelation/21-3.htm

https://www.biblemoneymatters.com/bible-verses-about-faith/

https://biblehub.com/revelation/2-7.htm

https://www.youtube.com/watch?v=Wd8H21ZcJL8

https://biblehub.com/revelation/21-4.htm

https://biblehub.com/revelation/21-3.htm

https://www.biblemoneymatters.com/bible-verses-about-faith/

https://biblehub.com/revelation/2-7.htm

https://www.youtube.com/watch?v=Wd8H21ZcJL8

https://bible.knowing-jesus.com/Hebrews/11/1

https://www.crosswalk.com/devotionals/girlfriends/girlfriends-in-god-january-2-2019.html

https://biblehub.com/1_corinthians/12-11.htm

https://biblehub.com/romans/15-30.htm

https://biblehub.com/romans/8-26.htm

https://biblehub.com/romans/8-27.htm

https://biblehub.com/john/15-26.htm

https://biblehub.com/acts/13-2.htm

https://www.youtube.com/channel/UCjrTpHHTTPF5gmsubJKIheA

https://www.youtube.com/watch?v=S5b9PSdl9xc

https://www.youtube.com/watch?v=wk2y4eu37_s

https://answers.yahoo.com/question/index?qid=20081119220619AAxd7kV

https://www.gotquestions.org/personal-Savior.html

https://answers.yahoo.com/question/index?qid=20110902144223AAkKwCS

http://rgsspeechless.blogspot.com/2010/04/many-love-letters-to-michael-1.html

https://decisioninlife.wordpress.com/2016/11/

https://www.gotquestions.org/eternal-life.html

https://answers.yahoo.com/question/index?qid=20060911012544AA7QMfr

http://embryhills.us/wp-content/uploads/2009/08/names-of-god.pdf

https://wordsearchbible.lifeway.com/products/35002-the-names-of-god/sample_text

https://answers.yahoo.com/question/index?qid=20141116075504AAlNIJL

Credit for Science & Bible Converge

https://www.biblegateway.com/passage?search=Genesis+15&version=NIV

http://www.bibletrack.org/cgi-bin/bible.pl?incr=0&mo=8&dy=29

http://embryhills.us/wp-content/uploads/2009/08/names-of-god.pdf

https://en.wikipedia.org/wiki/Trinity

https://biblehub.com/psalms/92-12.htm

https://biblehub.com/psalms/92-13.htm

https://biblehub.com/m/john/3-1.htm

https://biblehub.com/john/3-2.htm

https://biblehub.com/john/3-3.htm

https://www.biblegateway.com/cgi-bin/bible?search=John+3:1-16

https://biblehub.com/john/3-4.htm

https://www.coursehero.com/file/9909817/Argument-against-Christianity/

https://biblehub.com/ephesians/2-8.htm

https://www.youtube.com/watch?v=cPr9mNYYINE

https://www.biblestudytools.com/mark/16-16-compare.html

https://carm.org/should-we-baptize-name-father-son-and-holy-spirit-or-name-jesus

Credit for Science & Bible Converge

https://www.kingjamesbibleonline.org/Ephesians-2-8/

https://medium.com/@ufuomaee/think-biggere0a8c878beb0

https://en.islamway.net/article/8153/how-many-zsons-does-god-have

http://www.tucsonrevolution.com/baptism

https://biblehub.com/romans/6-4.htm

https://studybible.info/KJV/Ephesians%202:8

https://biblehub.com/matthew/10-32.htm

https://www.gotquestions.org/from-glory-to-glory.html

https://biblehub.com/nasb/2_corinthians/3-18.htm

Credit for Science & Bible Converge

https://www.youtube.com/watch?v=kxx8qAjTJY8

https://biblehub.com/2_corinthians/4-6.htm

https://veryfatoldman.blogspot.com/2017/08/reflection-how-to-consecrate-our.html

https://www.gotquestions.org/from-glory-to-glory.html

https://www.gotquestions.org/new-creation.html

https://www.gotquestions.org/intimacy-with-God.html

https://biblehub.com/proverbs/3-32.htm

https://biblehub.com/james/4-8.htm

https://www.biblegateway.com/passage/?search=James+4%3A8&version=NKJV

https://gregmclogan.com/oneness/

https://biblehub.com/nlt/luke/22-42.htm

https://wol.jw.org/en/wol/d/r1/lp-e/2005003

https://biblehub.com/interlinear/james/1-6.htm

https://biblehub.com/james/1-8.htm

https://biblehub.com/james/1-7.htm

https://biblehub.com/romans/7-21.htm

https://biblehub.com/romans/7-23.htm

Credit for Science & Bible Converge

https://biblehub.com/romans/7-24.htm

https://biblehub.com/romans/7-25.htm

https://www.facebook.com/Imday7/posts/?ref=page_internal

http://changebiblechurch.blogspot.com/2010/05/how-old-are-you.html

https://biblehub.com/romans/8-2.htm

https://biblehub.com/romans/8-3.htm

https://biblehub.com/ephesians/2-16.htm

https://biblehub.com/philippians/2-8.htm

https://biblehub.com/isaiah/42-1.htm

https://biblehub.com/matthew/20-28.htm

https://biblehub.com/john/1-14.htm

https://biblehub.com/romans/8-3.htm

https://biblehub.com/2_corinthians/8-9.htm

https://biblehub.com/2_corinthians/13-4.htm

https://biblehub.com/galatians/4-4.htm

https://biblehub.com/crossref/1_john/2-2.htm

https://biblehub.com/hebrews/2-17.htm

https://biblehub.com/philippians/2-7.htm

https://biblia.com/bible/niv/revelation/22/8-9

https://biblehub.com/m/romans/13-14.htm

https://www.biblegateway.com/passage/?search=rev+22&version=NASB

https://biblehub.com/revelation/22-8.htm

https://biblehub.com/revelation/22-9.htm

https://www.bible.com/bible/111/PHP.2.12.NIV

https://www.bible.com/bible/111/ROM.10.10.NIV

https://www.kingjamesbibleonline.org/John-5-39/

http://christianvanguardng.blogspot.com/2018/01/how-to-obtain-freedom-from-bondage.html

https://www.biblegateway.com/verse/en/1%20John%203%3A8

https://biblehub.com/revelation/21-1.htm

http://www.thebibleistheotherside.org/currentarticlep26p6.htm

https://biblehub.com/matthew/24-35.htm

https://sentence.yourdictionary.com/simply

https://biblehub.com/m/revelation/12-11.htm

https://www.christiancourier.com/articles/1105-what-is-the-morning-star-of-revelation-2-28

https://www.kingjamesbibleonline.org/Revelation-22-16/

https://biblehub.com/revelation/22-16.htm

https://biblehub.com/revelation/2-25.htm

https://biblehub.com/revelation/2-26.htm

https://biblehub.com/revelation/2-27.htm

https://blog.logos.com/2017/11/god-wasnt-alone-created-world-says-bible/

https://www.biblestudytools.com/leb/job/38-4.html

https://bnonn.com/what-is-the-kingdom-of-god-3/

Credit for Science & Bible Converge

https://biblehub.com/job/38-7.htm

https://biblehub.com/1_corinthians/15-21.htm

https://biblehub.com/1_corinthians/15-22.htm

https://biblehub.com/psalms/2-8.htm

https://www.christiancourier.com/articles/1105-what-is-the-morning-star-of-revelation-2-28

https://vod.whitedoveministries.org/wp-content/uploads/Webinar-69-notes-Ruling-Reigning.pdf

https://www.biblestudytools.com/1-corinthians/15-41-compare.html

https://biblehub.com/revelation/2-28.htm

https://en.wikipedia.org/wiki/Morning_Star

https://www.kingjamesbibleonline.org/Matthew-2-10/

https://biblehub.com/matthew/2-2.htm

https://www.kingjamesbibleonline.org/Matthew-2-2/

https://spaceplace.nasa.gov/review/dr-marc-space/solar-systems-in-galaxy.html

http://thingsworthknowin.blogspot.com/2016/03/how-many-solar-systems-are-in-our-galaxy.html

https://www.bible.com/bible/59/EXO.19.6.ESV

https://biblehub.com/1_peter/2-9.htm

https://biblehub.com/isaiah/61-6.htm

https://www.bible.com/bible/59/ISA.61.6.ESV

https://www.kingjamesbibleonline.org/Revelation-9-1/

https://www.biblestudytools.com/genesis/15-5.html

https://spaceplace.nasa.gov/review/dr-marc-space/solar-systems-in-galaxy.html

http://thingsworthknowin.blogspot.com/2016/03/how-many-solar-systems-are-in-our-galaxy.html

https://imgur.com/gallery/GtTJp7p

https://www.youtube.com/watch?v=ePiYFlaC28g

https://biblehub.com/psalms/19-1.htm

https://web.mit.edu/jywang/www/cef/Bible/NIV/NIV_Bible/PS+19.html

https://biblehub.com/isaiah/65-17.htm

https://biblehub.com/2_peter/3-13.htm

https://biblehub.com/mark/16-16.htm

https://biblehub.com/romans/14-23.htm

https://bible.knowing-jesus.com/2-Thessalonians/2/12

https://biblehub.com/proverbs/8-23.htm

https://biblehub.com/proverbs/8-22.htm

https://www.prayertents.com/b?search=Numbers+24:17&version=6

https://biblehub.com/numbers/24-17.htm

https://www.biblegateway.com/passage/?search=2+Peter+3%3A13&version=NIV

https://biblehub.com/2_corinthians/5-21.htm

https://biblehub.com/luke/22-44.htm

https://biblehub.com/romans/5-18.htm

https://whoishewordpresscom.wordpress.com/2016/03/26/minister-of-the-sanctuary/

https://www.gotquestions.org/I-never-knew-you-depart-from-me.html

https://biblehub.com/hebrews/8-1.htm

Credit for Science & Bible Converge

https://escapeallthesethings.com/depart-from-me-i-never-knew-you-meaning/

https://biblehub.com/par/matthew/10-33.htm

https://bible.org/seriespage/70-garden-gethsemane-luke-2239-46

https://biblia.com/bible/niv2011/luke/22/43-44

https://biblehub.com/m/luke/22-44.htm

https://dmstunt.wordpress.com/

https://vigilantcitizenforums.com/threads/there-should-be-peace-between-muslims-and- christians.5199/page-3

https://bible.org/question/what-does-greek-word-tetelestai-mean

https://preceptaustin.wordpress.com/2013/04/05/tetelestai-it-is-finished-paid-in-full/

https://biblehub.com/luke/22-44.htm

https://www.facebook.com/scfsrotterdam/posts

https://www.facebook.com/pages/category/Organization/Preceptaustin-136298009722380/

https://biblehub.com/matthew/27-46.htm

https://www.whatchristianswanttoknow.com/17-awesome-kjv-king-james-verse-bible-verses-about- heaven/

https://www.jesus-is-savior.com/Hells_Truth/forever.htm

Credit for Science & Bible Converge

https://www.kingjamesbibleonline.org/John-3-20/

https://www.biblestudytools.com/ecclesiastes/7-29-compare.html

https://biblehub.com/james/1-14.htm

https://biblehub.com/james/1-15.htm

https://biblehub.com/james/1-17.htm

https://biblehub.com/romans/5-11.htm

https://biblehub.com/romans/5-12.htm

https://biblehub.com/romans/5-13.htm

https://znanija.com/task/8356647

https://biblehub.com/matthew/25-41.htm

https://answers.yahoo.com/question/index?qid=20100713095728AAO7s2U

https://context.reverso.net/translation/english-spanish/do+not+know+what+they

https://www.desiringgod.org/messages/the-kingdom-of-god-is-righteousness-and-peace-and-joy-in-the- holy-spirit

https://biblehub.com/revelation/12-11.htm

https://answers.yahoo.com/question/index?qid=20110330201307AApvWdD

https://www.biblestudytools.com/acts/4-12.html

https://www.youtube.com/watch?v=udT-Ejqwi84

https://everydayexiles.com/2015/10/13/the-work-god-wants-from-you/tylerspeegle/

Credit for Science & Bible Converge

https://www.grc.nasa.gov/www/k-12/WindTunnel/Activities/first2nd_lawsf_motion.html

https://biblehub.com/titus/2-12.htm

https://www.bible.com/bible/111/ROM.7.19.NIV

https://www.sermonindex.net/modules/newbb/viewtopic.php?topic_id=13034&forum=36

https://www.youtube.com/watch?v=UDRFxdFAM-E

https://www.road2thecross.org/truth-blog/2015/12/7/the-jesus-way

https://www.biblegateway.com/verse/en/Deuteronomy%2031%3A13

https://biblehub.com/joshua/1-8.htm

https://www.biblestudytools.com/daniel/9-12-compare.html

http://leblogdelabergerie.com/prayers/Roman12.htm

https://www.biblegateway.com/passage/?search=1+Corinthians+12%3A12&version=ESV

https://biblehub.com/catholic/psalms/133-2.htm

https://bible.knowing-jesus.com/1-Corinthians/12/13

https://www.nationsford.org/resources/baptism

https://biblehub.com/acts/9-17.htm

https://bible.knowing-jesus.com/topics/Being-Filled-With-The-Spirit

https://www.quora.com/What-does-Jesus-mean-by-if-you-love-me-keep-my-commandments

Credit for Science & Bible Converge

https://answers.yahoo.com/question/index?qid=20080207184300AAcUOxC

http://anglicanboy.blogspot.com/2009/08/fullness-of-god-part-i.html

https://bible.knowing-jesus.com/2-Timothy/2/20

https://biblehub.com/1_corinthians/14-2.htm

https://bibledebates.wordpress.com/2017/10/05/god-gives-the-holy-spirit-to-those-that-ask-him/

https://biblehub.com/john/4-24.htm

https://biblehub.com/acts/8-31.htm

https://bible.knowing-jesus.com/Acts/8/32

https://www.youtube.com/watch?v=Wk0WgKzACt4

https://www.biblestudytools.com/acts/8-34.html

https://bible.knowing-jesus.com/Acts/8/35

https://www.desiringgod.org/messages/go-and-make-disciples-baptizing-them

https://bible.knowing-jesus.com/Isaiah/28/12

https://answers.yahoo.com/question/index?qid=20100115055710AA6cLj9

https://www.biblegateway.com/verse/en/Romans%201%3A16

https://www.biblestudytools.com/2-corinthians/8-19-compare.html

https://bible.knowing-jesus.com/Matthew/24/14

https://www.desiringgod.org/messages/the-limits-of-submission-to-man

https://www.biblegateway.com/verse/en/Romans%201%3A16

https://www.biblestudytools.com/2-corinthians/8-19-compare.html

https://bible.knowing-jesus.com/Matthew/24/14

https://www.desiringgod.org/messages/the-limits-of-submission-to-man

https://www.emeryhorvath.com/006-are-you-willing-to-give-jesus-everything-podcast/

https://biblehub.com/matthew/10-39.htm

https://nebraskafca.org/part-3-participation-in-sports-must-be-done-as-worship-to-god/

http://poteaucoc.org/From%20the%20Ministers%20Desk/The%20Prayers%20of%20the%20Holy%20Spirit.pdf

http://jdangelo6.blogspot.com/2014/05/

http://lionsandtigersandbears.blogspot.com/2010/01/will-people-call-you-thomas.html

https://www.bible.com/uz/bible/1/1PE.3.11-12

https://www.desiringgod.org/messages/battling-the-unbelief-of-covetousness

https://biblehub.com/2_corinthians/1-7.htm

https://biblehub.com/exodus/13-17.htm

https://enduringword.com/bible-commentary/exodus-13/

https://lavistachurchofchrist.org/LVanswers/2006/01-26.htm

https://dailyverses.net/hebrews/7/25

Prayer Request:

Made in the USA
Monee, IL
05 November 2026

67845310R00098